CLIMATE CHANGE AND INSURANCE

**CHRISTINA M. CARROLL, J. RANDOLPH EVANS,
LINDENE E. PATTON, & JOANNE L. ZIMOLZAK**

Cover by Elmarie Jara/ABA Publishing.
Cover photo by Jim Hamel.

Disclaimer: The opinions expressed are those of the author or the co-authors and do not reflect the opinions of Zurich Insurance Group Ltd., or any of its subsidiaries or affiliates, or McKenna Long & Aldridge LLP, or clients, or of the American Bar Association or the Tort Trial & Insurance Practice Section unless adopted pursuant to the bylaws of the Association.

© 2012 American Bar Association. All rights reserved.

No part of this publication may be reproduced, stored in a retrieval system, or transmitted in any form or by any means, electronic, mechanical, photocopying, recording, or otherwise, without the prior written permission of the publisher. For permission, contact the ABA Copyrights and Contracts Department by e-mail at copyright@abanet.org or fax at 312-988-6030, or complete the online request form at http://www.americanbar.org/utility/reprint.html.

Printed in the United States of America

16 15 14 13 12   5 4 3 2 1

Library of Congress Cataloging-in-Publication Data

Carroll, Christina M.
  Climate Change and Insurance / by Christina M. Carroll.—1st ed.
     p. cm.
  Includes bibliographical references and index.
  ISBN 978-1-61438-733-6 (print : alk. paper)
  1. Climatic changes—Law and legislation—United States. 2. Climate change insurance—Law and legislation—United States. 3. Liability for environmental damages—United States. I. Title.

KF3783.C37 2012
346.73'086—dc23

2012039222

Discounts are available for books ordered in bulk. Special consideration is given to state bars, CLE programs, and other bar-related organizations. Inquire at Book Publishing, ABA Publishing, American Bar Association, 321 North Clark Street, Chicago, Illinois 60654-7598.

www.ShopABA.org

*Chief Climate Product Officer, Zurich Insurance Group Ltd. Opinions expressed are those of the author or the co-author and do not necessarily reflect the opinion of Zurich Insurance Group Ltd. or any of its subsidiaries or affiliates.

CONTENTS

About the Authors ix
Acknowledgments xiii

CHAPTER 1: Climate Change and Insurance: An Overview 1
    A. Potential Exposures 2
        1. Potential Enterprise Liability 2
        2. Potential Claims Liability 3
    B. Implications by Coverage Type 4
    C. Risk Management and Product Development Considerations 8
        1. Potential Enterprise Liability 8
        2. Potential Claims Liability 9

CHAPTER 2: Climate Change, Global Warming, and Greenhouse Gases: Background 11
    A. Defining "Global Warming" and "Climate Change" 11
    B. The Importance of Definitions in the Insurance Context 13
    C. Greenhouse Gases and the Causation Issue 14
    D. Greenhouse Gases (GHGs) 16
    E. Conclusion 20

CHAPTER 3: Climate Change as an Emerging Risk and Comparison to Historic Emerging Risks 21
    A. Climate Change as an Emerging Risk 21
    B. Evolution of Other Emerging Risks 22

CHAPTER 4: The Likely Plaintiffs and Targets of Emerging Climate Change-Related Litigation and Regulation 25
    A. Potential Impacts and Likely Claimants 25
    B. Targets/Defendants 26
        1. Emitters of GHGs 26
        2. Service Providers to Targeted Emitters 30
        3. Impact to Insurers 31

CHAPTER 5: Existing and Emerging Climate Change-Related Regulation ... 33
A. The International Backdrop ... 34
B. National Regulation ... 38
  1. EPA Action Pursuant to the Clean Air Act ... 38
     a. *Massachusetts v. EPA* ... 39
     b. EPA Endangerment and Cause or Contribute Findings for GHGs ... 40
  2. EPA and NHTSA Light-Duty Vehicle GHG Standards and Corporate Average Fuel Economy Standards ("Tailpipe Rule") ... 43
  3. EPA Timing Rule ... 44
  4. EPA's GHG Reporting Rule ... 45
  5. EPA Prevention of Significant Deterioration and Title V Greenhouse Gas Tailoring Rule ... 46
  6. Standards of Performance for Carbon Dioxide Emissions for New Stationary Sources: Electric Utility Generating Units ... 50
  7. Potential Impact of Emerging Clean Air Act Regulation on Climate Change-Related Litigation ... 51
C. Congressional Action ... 51
D. Regulation and Potential Litigation Related to Carbon Storage ... 52
E. Renewable Energy: An Overview ... 53
F. State and Local Regulation ... 56
G. Regional Agreements ... 59
H. State Insurance Regulation ... 61
  1. The National Association of Insurance Commissioners (NAIC) ... 61
     a. NAIC Climate Risk Disclosure Survey ... 62
     b. 2011 Ceres Report on 2010 NAIC Survey Results ... 63
     c. Recent NAIC Activities ... 65
     d. California, New York, and Washington Mandatory Survey ... 66
  2. California Green Insurance Act of 2010 ... 66
  3. Pay-As-You-Drive Regulations ... 67

CHAPTER 6: Climate Change-Related Litigation ... 69
A. Common Law Tort Litigation ... 70
  1. Climate Change-Related Tort Litigation: An Overview ... 70
  2. *American Electric Power Co. v. Connecticut* ... 71
  3. *Kivalina v. ExxonMobil Corp.* ... 76
  4. *California v. General Motors Corp.* ... 80

	5. Tort Litigation Related to Episodic Climatic Events 81	
	a. *Turner v. Murphy Oil USA, Inc.* 81	
	b. *Comer v. Murphy Oil USA, Inc.* 82	
	6. Future Trends: Climate Change-Related Tort Litigation 87	

5. Tort Litigation Related to Episodic Climatic Events 81
   a. *Turner v. Murphy Oil USA, Inc.* 81
   b. *Comer v. Murphy Oil USA, Inc.* 82
6. Future Trends: Climate Change-Related Tort Litigation 87
   a. Additional Types of Claims Including Products Liability 87
   b. The Causation Issue 88
   c. Foreseeability 91

B. Climate Change-Related Insurance Coverage Litigation:
   An Introduction to *AES Corp. v. Steadfast Insurance Co.* 92
C. Public Trust Litigation 93
D. Claims for Natural Resource Damages Pursuant to CERCLA 95
E. Claims Arising out of Corporate Disclosure and Management
   of Climate Change Risk 97
   1. Shareholder Resolutions Related to Climate Change 98
   2. Potential Litigation Related to Misrepresentation, Concealment,
      or Mismanagement of Climate Change Risk 102
   3. Role of the SEC in Climate Risk Disclosure 102
      a. Background: Petitions to the SEC for Further Regulation
         of Climate Change Risk Disclosure 102
      b. Current Securities Laws, Regulations, and Guidance
         Requiring Disclosure of Climate-Change Issues 104
   4. Climate Change-Related Activity in New York Pursuant
      to the Martin Act 106
   5. Voluntary Climate Change Disclosure Programs 108
      a. Carbon Disclosure Project 108
      b. Principles for Responsible Investment 110
      c. Dow Jones Sustainability Index 111
      d. FTSE4Good Index Series 112
      e. The Climate Registry 114
      f. Periodic Stakeholder Reporting (Annual Reports) 115
F. Potential Litigation Arising out of Efforts to Address Climate
   Change Issues 115
   1. Utilization of the National Environmental Policy Act (NEPA)
      and State Environmental Policy Acts (SEPAs) to Address
      Climate Change 115
      a. National Environmental Policy Act (NEPA) 115
      b. State Environmental Policy Acts (SEPAs) 117
      c. Implications 119

  2. Disputes Arising out of Carbon and GHG Markets 120
  3. Greenwashing Litigation.. 122
   a. Greenwashing and Consumer Actions Related
    to Misleading or False Advertising......................... 123
   b. Federal Trade Commission Focus on "Green" Marketing 128
  4. Green Building Litigation...................................... 129
   a. Green Building Standards.................................. 130
   b. Litigation Resulting from Governmental Legislation
    and Regulation 131
   c. Contractual and Other Disputes 132
 G. Potential GHG-Related Enforcement Actions 134

CHAPTER 7: Intersection of Climate Change and Insurance 135

 A. Existing and Emerging Climate Change-Related Exposure
  Arising out of First- and Third-Party Coverage of Insureds
  ("Potential Claims Liability") 135
  1. Overview of Potential Claims Liability.......................... 135
  2. First-Party Coverage 136
   a. Implications for Property Coverage 136
   b. Implications for Property (Time Element) Coverage 136
  3. Third-Party Coverage 138
   a. Implications for D&O Coverage 138
   b. Implications for Commercial General Liability Coverage....... 141
   c. Implications for Environmental Liability Coverage............ 143
   d. Implications for Professional Liability/E&O Coverage 144
   e. Implications for Contractor's Protective Professional
    Indemnity and Liability Insurance and Owner's Protective
    Professional Indemnity Insurance 145
  4. Potential for Bad Faith Claims.................................. 146
   a. Bad Faith Claims Based on Inconsistent Position on Climate
    Change-Related Issues..................................... 146
   b. Bad Faith Claims Based on Denial of Coverage................ 146
 B. Emerging Climate Change-Related Exposure: Potential
  Enterprise Liability.. 147
  1. Potential Liability Based on Corporate Disclosures 147
  2. Potential Liability Related to Provision of Loss Control
   Consulting Services.. 148

C. New Applications of Existing Coverages and New Products. 148
    1. Traditional Insurance Products for Renewable Energy Projects. 149
        a. Property Insurance. 149
        b. Boiler and Machinery Insurance. 150
        c. Commercial General Liability Insurance. 151
    2. Insurance Policies Tailored for the Renewable Energy Industry. 151

CHAPTER 8: Potential Claims Liability: Potential Exposure Drivers 153

A. Claims-Made and Reported 154
    1. The Claim Must Be Made During the Policy Period
       or Extended Reporting Period. 155
    2. The Retroactive Date Requirement. 156
B. Occurrence-Based. 156
    1. *AES Corp. v. Steadfast Insurance Co.* 156
    2. *Cinergy Corp. v. Associated Electric & Gas Insurance Services, Ltd.*:
       Damages Not Caused by An Occurrence. 159
    3. The Effect of the "Trigger" on Occurrence 160
        a. Exposure 161
        b. Manifestation 161
        c. Injury-in-Fact 162
        d. Continuous 163
    4. Long-Tail Claims 164
C. Pollution Exclusions and Grants of Pollution Coverage. 164
    1. Definition of "Pollutant". 164
        a. The Interpretation of the Term "Pollutant". 165
        b. Whether GHGs Are Pollutants. 167
    2. Pollution Exclusion 170
        a. The Three Types of Pollution Exclusions. 171
        b. Time Element, Limited-Pollution Coverage Endorsements. 177
        c. Special Issues Associated with Pollution Exclusions
           in the D&O Context 178
    3. Grants of Pollution Coverage. 179
D. Other Potentially Relevant Exclusions 180
    1. Bodily Injury and Property Damage Exclusion in D&O Policies. . . . 180
    2. Expected or Intended Injury Exclusion. 181
    3. Known Loss, Known Injury or Damage, Loss in Progress Clauses . . 183
    4. Preexisting Conditions Exclusion 185

        5. Injunctive Relief Exclusion 186
        6. Maintenance/Betterment Exclusions 187
        7. Punitive, Treble, Exemplary, or Multiple Damages Exclusion 187
        8. Fines, Penalties, and Taxes Exclusion 188
        9. Intentional/Criminal Act Exclusions 189
    E. Limitations 191
        1. Defense Within or Outside of Limits 191
        2. Batch Clauses/Multiple Claim Clauses 191
        3. Anti-Stacking Clauses 192
    F. General Conditions: Notice Requirement 193
    G. Will Exposure Drivers Result in Exposure? 194

Glossary of Acronyms & Terms 197
Index 225

ABOUT THE AUTHORS

Christina M. Carroll is a partner in McKenna Long & Aldridge LLP's Washington, D.C., office. Her diverse practice focuses primarily on complex litigation, including insurance coverage and bad faith, environmental, toxic tort, and federal preemption litigation. Additionally, she counsels clients on the risks and opportunities associated with climate change.

Ms. Carroll represents a variety of clients in the insurance, chemical, healthcare, and technology industries. In the insurance realm, she represents insurance companies in coverage and bad faith disputes and also counsels them on strategic issues including emerging risks and new product development. Ms. Carroll has handled matters involving directors and officers, property, commercial general liability, professional liability, and automobile liability insurance. She brings her environmental and scientific background together with her insight in insurance-related law to provide a unique skill set in developing strategies for monitoring and addressing climate change, hydraulic fracturing, green building, greenwashing, cybersecurity, and other emerging risks.

Ms. Carroll is a frequent lecturer and author on the subject of climate change and the intersection between climate change and insurance. In addition to her role as a practitioner, Ms. Carroll is the firm's Hiring Partner for the Washington, D.C., office.

J. Randolph "Randy" Evans is a partner in the Atlanta and Washington, D.C., offices of McKenna Long & Aldridge LLP. He is the chair of the firm's Financial Institutions practice. He handles high-profile, complex litigation matters in state and federal courts throughout the United States for some of the largest companies in the world.

In addition, Mr. Evans served as outside counsel to the Speakers of the 104th–109th Congresses of the United States, Dennis Hastert and Newt Gingrich. He also represents several current Members and former Members of Congress, as well as elected officials in Georgia.

Mr. Evans advises clients with respect to the emerging risk of climate change, including with regard to the likely trajectory of climate change-related litigation. In this regard, Mr. Evans has catalogued and developed a unique construct for evaluating emerging risks—the "five phases of mass tort recoveries"—as published in "Is Past Prologue to Climate Change Liability?" *Law360*, New York (May 31, 2011) and other publications. He also advises clients on corporate governance issues. Mr. Evans is a frequent lecturer and author on the subjects of climate change, professional liability and ethics, and government ethics and politics.

Mr. Evans has been named in *Best Lawyers in America* in the practice areas of Commercial Litigation and Legal Malpractice Law.

Lindene E. Patton is Chief Climate Product Officer for Zurich Insurance Group (Zurich). She is responsible for product development and risk management related to climate change.

Ms. Patton is a member of the World Economic Forum Global Advisory Council on Measuring Sustainability. She serves as the Vice-Chair of the Climate Change and Tort Liability Sub-Committee of the Geneva Association. She is a member of the Advisory Council to the Resources for the Future's Center for the Management of Ecological Wealth (RFF's CMEW). Ms. Patton serves on numerous government and non-governmental advisory boards, including the Executive Secretariat of the U.S. National Climate Assessment Development and Advisory Committee and the U.S. Environmental Protection Agency (EPA) Environmental Financial Advisory Board. She is an advisory board member for the University of California at Santa Barbara's Bren School of Environmental Science and Management and is a member of the ICLEI for Sustainable Governments Adaptation Experts Advisory Committee. Ms. Patton serves as an Advisory Board Member to the Bloomberg monthly publication, *The Environmental Due Diligence Guide*, and the U.S. EPA Environmental Technology Verification Program.

Ms. Patton is an attorney licensed in California and the District of Columbia and an American Board of Industrial Hygiene Certified Industrial Hygienist. She holds a Bachelor of Science in biochemistry from the University of California, Davis, a Master of Public Health from the University of California, Berkeley, and a Juris Doctor from Santa Clara University School of Law.

Joanne L. Zimolzak is a partner in McKenna Long & Aldridge LLP's Washington, D.C., office. Ms. Zimolzak's practice focuses on business litigation relating to commercial and government contracts, including insurance coverage and bad faith, contract disputes, Administrative Procedure Act and Freedom of Information Act matters, and fraud/compliance matters. She also advises clients about climate change-related risks and opportunities for various business sectors.

Ms. Zimolzak represents a variety of clients in the insurance, aerospace/defense, energy, and information technology industries. The insurance component of her practice includes representing multi-national insurance companies in all types of coverage disputes, with emphasis on matters involving bad faith allegations and significant potential exposures. A current Vice-Chair of the ABA's Tort Trial and Insurance Practice Insurance Coverage Litigation Committee, Ms. Zimolzak also routinely counsels insurers concerning coverage and risk management issues, emerging liability risks, new product development, regulatory compliance, and other aspects of their business.

Ms. Zimolzak is a frequent lecturer, author, and contributor on the subject of climate change. Her work has been featured in *Business Insurance*, *Climate Change Business Journal*, and *Tort Trial & Insurance Practice Law Journal*, among other publications. In addition to her role as a practitioner, Ms. Zimolzak serves as the Office Managing Partner of the firm's Washington, D.C., office and leads the Washington-based component of the firm's Financial Institutions practice.

ACKNOWLEDGMENTS

The authors wish to thank the many people who helped them produce *Climate Change and Insurance*. This book is the first of its kind, and thus the authors are thankful to all who were involved in the project.

We thank our colleagues in the McKenna Long & Aldridge LLP Climate Change, Energy and Sustainability practice group and the Litigation Department for their support. In particular, we would like to acknowledge Kristin Landis, Esq. of McKenna Long & Aldridge for her commitment and great work on this project. She drafted significant portions of Chapters 7 and 8 and assisted with other aspects of the book. Her assistance was invaluable. In many respects, she is a co-author.

Other attorneys, paralegals, and staff also contributed to the work. They include McKenna Long & Aldridge attorneys Jane Moffat and Andrew Shaw, and former McKenna Long & Aldridge associate Michael Baumrind. Paralegal Carol Wagner also has provided tremendous assistance with the book. Legal secretary Susan Valentine tirelessly read and corrected numerous drafts of this work. Martha Klein and Douglas Malerba, McKenna Long & Aldridge librarians, both provided timely updates and excellent research. Numerous other members of MLA's secretarial and paralegal team also provided careful assistance. Many thanks to everyone who helped make this book a reality.

Christina M. Carroll, Esq.
J. Randolph Evans, Esq.
Lindene E. Patton, Esq.
Joanne L. Zimolzak, Esq.

CHAPTER 1

Climate Change and Insurance: An Overview

Climate change has become a leading issue in scientific, political, and legal circles, and this may have profound implications for the insurance industry. More frequent catastrophic events such as floods, droughts, fires, hurricanes, tornadoes, monsoons, and sandstorms, whether or not attributable to climate change, challenge the insurance industry's abilities to measure, predict, and price risk in the first-party context. At the time insurers underwrote many third-party policies currently in force and historic occurrence policies that could still create exposure, neither insurers nor insureds focused on the potential impacts of "climate change" as we understand and discuss that concept today. Thus, climate-change risk probably was not contemplated in developing policy language, underwriting the risk, or developing premium models. The potential exposure associated with climate-change impacts, however, is now beginning to emerge. Tort liability arising out of episodic climatic events targeting operators of facilities and manufacturers with no physical presence in the geographic location of the event indicates that historical insurer management practices, such as limiting aggregate exposures by geographic location, may not be sufficient to manage financial liability. Risks independent of episodic climatic events also are emerging and require additional risk-management approaches.

Climate change is consistently referred to as an "emerging risk" because stakeholders, including potentially regulated entities like targeted emitters and insurers, do not yet know how courts will treat climate change or what mechanisms policymakers will create to further mitigate or adapt to climate change-related risks. Mitigation focuses on interventions to reduce the sources or enhance the "sinks" for greenhouses gases (GHGs).[1] Adaptation refers to adjustments in natural or human systems in response to actual or expected climate-change impacts, which moderate harm or exploit beneficial opportunities.[2]

---

1. *See, e.g.*, United Nations Framework Convention on Climate Change, Adaptation, Private Sector Initiative (PSI) 2012, *available at* http://unfccc.int/adaptation/nairobi_work_programme/private_sector_initiative/items/4623.php. A GHG "sink" is anything that absorbs more GHGs than it releases.
2. *Id.*

In addition, the science used by courts and policymakers is still evolving and its interpretation is often politicized. As with all emerging issues, impacted parties cannot definitively calculate future costs when parameters of the risk, such as frequency and severity, have no specific history. Recent events make apparent, however, that this risk has emerged.

Historical experience with analogous emerging risks (e.g., asbestos, tobacco, and CERCLA/Superfund) indicates that no single change will provide a "silver bullet" solution for managing the risk or securing new opportunities. Risk management and product development strategies could include (1) policy language changes, (2) management of individual policy limits of liability and term lengths, (3) management of overall aggregate liability exposures for a given class or line of business, (4) creation of intentional coverage grants to serve market needs, or (5) participation in the public policy debate regarding changes in applicable law and creation of new legal standards in order to secure the ability to continue to underwrite in a field with greater legal predictability.

A. Potential Exposures

Despite the inherent uncertainty associated with an emerging risk, two categories of potential climate change-related core exposures are apparent: (1) potential exposures arising out of insurer institutional statements and activities ("potential enterprise liability") and (2) potential exposures arising out of relationships with insureds ("potential claims liability") with respect to risk management products and services. The first potential core exposure arises from such corporate functions as climate change-related disclosures and development of knowledge of climate change-related issues through underwriting, claims processing, and fee-for-service consulting work. The second potential core exposure develops when insureds faced with climate change-related claims look to their existing insurance policies for coverage.

1. Potential Enterprise Liability

Insurers could face claims as a result of their institutional statements and obligations like disclosures, data collection, and loss control and consulting fee-for-service work. Insurers have mandatory disclosure obligations and also voluntarily issue disclosures that could result in liability. They also have to respond to mandatory climate-change risk disclosure surveys in some states. It is possible that states will take regulatory action against insurers as a result of their responses. In the future, regulators could require further disclosure on proprietary matters and increase reserve requirements. Some insurers also participate in a number of voluntary programs requesting climate change-related information such as the Carbon Disclosure Project, the Dow Jones Sustainability Indexes, and FTSE4Good, and also publish their own material on climate change. While such participation is important for corporate social responsibility efforts, marketing, and profile raising, voluntary climate-change risk disclosures also carry some potential problems: they can be used by regulators, plaintiffs in bad faith actions trying to show inconsistent corporate positions, plaintiffs in other stakeholder litigation, or shareholders if the company is publicly traded.

Historical trends in asbestos litigation demonstrate that plaintiffs sometimes sue insurers directly if insurers had particularized knowledge of an emerging risk and other targets became insolvent. For example, as solvent asbestos defendants became increasingly rare, plaintiffs began targeting insurers directly, often relying on claims stemming from the insurers' failure to disclose information regarding the dangers of asbestos to an unknowing public. Through claims handling and dealings with insureds, insurers had developed vast knowledge of asbestos risk, and some claimed insurers had a duty to disclose this information. Generally, third-party asbestos claims against insurers have been unsuccessful, but litigation and nuisance settlement costs added up. A similar scenario could arise in relation to climate change as insurers respond to increasing numbers of climate change-related natural-catastrophe claims; promote green building in replacement coverage after episodic climatic events and consumers or policymakers demand use of green building materials to lower carbon footprints; and otherwise integrate sustainability principles into operations. By handling replacement claims and property, professional, or other claims related to climatic events; green-building performance; and other adaptive and mitigative measures, insurers may develop knowledge of the climatic science, safety of green materials and related information about adaptive and mitigation technologies employed to address risks of climate change. This could lead to claims against insurers arising out of their particularized knowledge of any of these issues.

2. Potential Claims Liability

Several types of climate change-related litigation are emerging that could result in potential claims to an insurer:

(1) tort or tort-related litigation including class actions or consolidated actions including:
  (a) tort claims arising out of episodic climatic events (e.g., hurricanes) such as *Turner v. Murphy Oil USA, Inc.* and *Comer v. Murphy Oil USA, Inc.* (see Chapter 6.A.5);
  (b) tort claims arising out of more gradual or projected alleged consequences of climate change such as *Kivalina v. ExxonMobil Corp.* (see Chapter 6.A.3);
  (c) tort claims due to state attorney general (AG) or nongovernmental organization (NGO) activism such as *American Electric Power Co., Inc. v. Connecticut* and *California v. General Motors* (see Chapters 6.A.2 and 6.A.4);

(d) product-liability claims alleging that products like cars and other products that emit GHGs are defective as designed; and

(e) natural resource damage claims based on CERCLA.

(2) regulatory enforcement actions related to GHG emissions (if GHG emissions limits are enacted);

(3) shareholder, Securities and Exchange Commission, or state actions related to misrepresentation, concealment, or mismanagement of climate change-related risk;

(4) a variety of types of claims against service providers arising out of efforts to fix climate change-related problems, such as claims related to preparation of environmental impact statements and mitigation strategies pursuant to federal and state environmental policy acts; claims related to preparation of carbon footprints; disputes related to emerging carbon markets; greenwashing litigation related to allegedly green products; green-building litigation; and failures of emerging technologies like carbon capture and sequestration (CCS) and alternative energy; and

(5) insurance coverage litigation associated with all of the above. Some of these claims may not ultimately succeed, but could prove costly for an insurer to defend.

B. Implications by Coverage Type

Policyholders faced with climate change-related claims may seek coverage pursuant to (1) third-party directors and officers (D&O), commercial general liability (CGL), environmental, and professional/errors and omissions (E&O) policies; (2) first-party property, accident, health, and political risk policies; and (3) first- and third-party policies like owners protective professional indemnity insurance (OPPI) and contractors protective professional indemnity and liability insurance (CPPI). Depending on the nature of the climate change-related claim, multiple types of coverage may be triggered simultaneously. For example, a court could construe a CGL pollution exclusion narrowly and conclude that GHGs are not pollutants but at the same time declare GHGs pollutants for purposes of a coverage grant in an environmental liability policy.

The potential implications for each general policy type are as follows:

- *Implications for D&O Coverage*—In recent years, shareholder resolutions related to climate change have been on the rise, and investors have demanded increased disclosure of climate change-related risk. Voluntary disclosure programs abound. The SEC has issued guidance on climate change disclosures and on shareholder resolutions, which impact climate change-related resolutions. In this political environment, directors and officers eventually may face class actions and/or shareholder derivative actions if a corporation, or its directors and officers, misrepresent, mismanage, or fail to disclose climate change-related risk. Such failures result in harm to a corporation, such as loss of revenues, loss of market share, falling stock price, reputational damage, and/or missed business opportunities related to climate change issues. Regulators also may bring actions against insureds related to faulty disclosures. Insureds facing claims related to concealment, misrepresentation, or mismanagement of climate change-related risk likely will seek reimbursement of defense costs and indemnification under D&O policies.

In analyzing available coverage, the pollution exclusion, intentional/criminal acts exclusion, definition of loss, limits of liability provision related to interrelated wrongful acts, bodily injury and property damage exclusions, and definition of claim provision, among others, should be considered. Certain insurers have entered the market with affirmative coverage for claims related to mismanagement of climate change or deleted exclusions (e.g., Zurich and Liberty Mutual).

- *Implications for CGL Coverage*—Insureds hit with climate change-related tort lawsuits probably will look to CGL policies for coverage because CGL insurance generally provides broad coverage for defense and indemnity of claims for bodily injury, personal injury, and property damage (potentially including natural resource damages). Insureds in the manufacturing and green building industry also may look for advertising injury liability coverage if they are subject to greenwashing claims. CGL product liability endorsements also may be implicated by new climate change-related claims. Although no climate change-related product-liability claims have been made thus far, it is possible that individuals could bring product-liability claims against an insured, such as failure to warn or design-defect claims related to products that emit GHGs. CGL is an attractive target because claims may implicate numerous occurrence-based policies and the policies generally provide defense coverage in addition to policy limits for any indemnity. Even if climate change-related cases do not ultimately succeed, defense costs could be high.

CGL policies have pollution exclusions, known loss or loss in progress exclusions and limitations, expected or intended exclusions, batch clauses, and failure to conform exclusions that may be relevant in the climate change context. Thus far, there has only been one climate change-related coverage dispute in the courts. In *AES Corp. v. Steadfast Insurance Co.*, the Virginia Supreme Court concluded that there can be no "occurrence" giving rise to coverage where the underlying complaint alleges that the defendant intentionally emitted GHGs and the alleged "natural and probable consequence" of such emissions is global warming and related damages.[3] Given it decided there was no occurrence, the Virginia court did not need to reach the other issues raised by Steadfast—the pollution exclusion and known loss argument. Other jurisdictions may be called on to decide climate change-related CGL coverage disputes involving these issues.

- ***Implications for Environmental Liability Coverage***—Environmental liability insurance generally covers damages and cleanup costs arising out of "pollution events" at, on, under, or migrating from covered locations and defense costs associated with such events. Thus, it is highly likely that insureds facing climate change-related claims will seek coverage under environmental policies. Whereas any conclusion that GHGs are pollutants for purposes of pollution exclusions could decrease insurers' exposure under CGL policies, a determination that GHGs are pollutants could increase insurers' exposure under environmental policies. Courts probably will consider recent EPA regulations related to GHGs in evaluating whether GHGs are pollutants for purposes of affirmative coverage grants in insurance policies. Thus, climate change presents a special risk in the environmental liability insurance field.

    Insureds may seek coverage under environmental liability policies for tort claims arising out of episodic environmental events, other common law tort claims (such as *AEP*, *Kivalina*, and *Comer*), and statutory environmental claims such as natural resource damages under CERCLA. In addition to tort lawsuits and CERCLA NRD actions, insureds may face statutory liability pursuant to the Clean Air Act or new GHG statutory schemes if ever enacted by Congress. For example, insureds may attempt to seek costs of compliance through their environmental liability policies. Betterment exclusions could bar exposures for day-to-day operational costs such as costs associated with Clean Air Act–related upgrades, but the specific wording of the particular exclusion should be evaluated and considered in context.

    The fact that most environmental policies are written on a claims-made rather than an occurrence basis will temper some of this risk because insureds will not be able to look to historic or multiple policies. Circumstance reporting of climate change events, however, may become more commonplace as

---

3. 725 S.E.2d 532 (2012).

insureds try to preserve their right to demand coverage for climate change events even under claims-made policies.

In evaluating exposure under environmental liability policies, the following policy provisions may be relevant: definition of "natural resource damages"; intentional/criminal acts exclusion; fines, penalties, and punitive damages exclusion; and limits of liability provisions for multiple claims arising out of the same pollution event, multiple insureds or claimants, multiple coverages, multiple policy periods, and sub-limits.

• *Implications for Professional/E&O Coverage*—Climate change-related claims could arise out of a wide variety of professional activities such as carbon footprinting, preparation of environmental impact statements and mitigation strategies, green building, and greenwashing. As EPA, states, and nonprofits demand disclosure of GHG emissions from corporations, professionals are being called on to account for or calculate GHG emissions for corporate clients in reports to EPA or other federal or state authorities. For example, regulated emitters will rely on professionals to help prepare reports pursuant to the EPA GHG Reporting Rule and in connection with preparation of environmental impact statements and mitigation strategies related to obtaining permits at the state and local level. If professionals make mistakes in the course of doing such work, their customers may file claims against them.

Professionals also could be subject to liability if problems arise from misrepresentations or mistakes made in connection with green building, installation of emissions-reduction solutions, and possibly handling of carbon credits. For example, if a professional fails to take the promised steps to mitigate GHGs in a green building project, causing a loss of LEED certification or loss of tax benefits, the client may have a claim against the professional. If a consultant assists a client in developing green products and the client is saddled with greenwashing claims, the client may in turn sue the consultant. As climate change progresses and adaptation solutions become increasingly important, claims arising out of adequacy of professional services related to resilience solutions and disaster management planning may also increase.

Professionals will seek coverage for these kinds of climate change-related claims. Whether or not exposure is significant may depend on whether the professional liability policies contain pollution or intentional/criminal acts exclusions. Exclusions for fines; penalties; taxes; punitive, exemplary, and multiple damages; and products liability also may be relevant considerations in the coverage analysis. In light of the likely emerging claims, it will be important to identify and monitor the evolving standard of care and to revamp underwriting criteria and pricing in this area.

- *Implications for Property, Accident, and Health Coverage*—Episodic climatic events will continue to generate first-party claims under personal and commercial property policies. These same events will also continue to generate accident and health-based claims. Regardless of their potential relation to climate change, insurers generally have systems in place to monitor how severe weather events are affecting the frequency and severity of first-party property claims. Many insurers and their experts model storm patterns and other natural phenomena without labeling this "climate change" work. This book does not focus on first-party claims in detail.
- *Implications for Political Risk Coverage*—Insureds may seek political risk coverage for carbon and other GHG emissions credit transactions. Political risk insurance may protect against the risk of a host government's actions that prevent an insured from receiving benefits associated with emission credits. It also might provide coverage for political violence, including war, riot, unauthorized repatriation, and terrorism, which might disrupt operations related to the credits. Some insurers have specifically decided to cover such risk through new product offerings in this area.
- *New Products*—Traditional property and casualty insurance, along with specific new products targeted to the needs of renewable and alternative energy projects, such as carbon capture and sequestration (CCS) or warranty policies, may assist owners, operators, and contractors with climate change-related risk management. Insurers may find that climate change creates a fertile ground for new applications of existing products and demand for new and innovative products to address emerging circumstances.

C. Risk Management and Product Development Considerations

1. Potential Enterprise Liability

Careful vetting of climate change disclosures can help reduce risk because litigants might try to use such communications to establish knowledge or an enterprise position even when the litigation or claims matter at issue is in a jurisdiction foreign to that where the disclosure or communication was made. Insurers also can try to avoid taking inconsistent positions in claims against their own insurance carriers and in coverage disputes with their insureds.

2. Potential Claims Liability

Given there have been few claims to date, there is no clear exposure driver. Nevertheless, the following appear to be particularly significant in the climate change context: (a) whether a policy contains a pollution exclusion or grant of pollution coverage, (b) whether a policy offers defense in or outside of limits, and (c) whether a policy is claims-made without circumstance reporting options or triggered by an occurrence. The range of potential exposure drivers are discussed in Chapter 8.

In addressing climate change-related risk, insurers could consider developing new products specifically addressing the risks unique to climate change (and thereby channeling emerging market need into products better suited to manage these emerging risks for both insurers and insureds); new exclusions; sub-limits; restrictions on defense outside of limits; restrictions on underwriting certain industries; development of underwriting guidelines for new industries and for changing exposures to existing industries, changes in pricing and risk profiling; strategic changes to language of existing exclusions, limitations, and other policy provisions; and strategic placement of existing exclusions, limitations, and other policy provisions in additional kinds of policies.

CHAPTER 2
Climate Change, Global Warming, and Greenhouse Gases: Background

A. Defining "Global Warming" and "Climate Change"

Most scientists agree that significant changes are occurring in the Earth's climate. Since the late 19th century, scientists note, the global average temperatures have increased about 0.74 degree Celsius (plus or minus 0.18 degree Celsius).[1] According to a National Research Council (NRC) study, the "global mean surface temperature was higher during the last few decades of the 20th century than during any comparable period in the preceding four centuries."[2] The NRC study also estimates, based on proxy data, that temperatures at many, but not all, locations studied were higher during the past 25 years than during any period of comparable length since 900 A.D.[3] That conclusion is more uncertain than the prior one.[4] The Intergovernmental Panel on Climate Change (IPCC), a scientific body for the assessment of climate change established in

---

1. Kevin E. Trenberth et al., *Observations: Surface and Atmospheric Climate Change*, in CLIMATE CHANGE 2007: THE PHYSICAL SCIENCE BASIS. CONTRIBUTION OF WORKING GROUP I TO THE FOURTH ASSESSMENT REPORT OF THE INTERGOVERNMENTAL PANEL ON CLIMATE CHANGE 237 (Solomon et al. eds., 2007); *see also* National Oceanic & Atmospheric Administration, National Climatic Data Center, *Global Warming Frequently Asked Questions*, http://www.ncdc.noaa.gov/oa/climate/globalwarming.html (last updated Aug. 21, 2012).

2. NATIONAL RESEARCH COUNCIL (NRC), SURFACE TEMPERATURE RECONSTRUCTIONS FOR THE LAST 2,000 YEARS 3 (2006); *see also* EPA, TECHNICAL SUPPORT DOCUMENT FOR ENDANGERMENT AND CAUSE OR CONTRIBUTE FINDINGS FOR GREENHOUSE GASES UNDER SECTION 202(A) OF THE CLEAN AIR ACT 31 (Dec. 7, 2009), *available at* http://www.epa.gov/climatechange/Downloads/endangerment/Endangerment_TSD.pdf (discussing NRC study). Similar findings have been echoed in other more recent studies. *See, e.g.*, D.S. Arndt, M.O. Baringer & M.R. Johnson, eds., *2010: State of the Climate in 2009*, 91 BULL. OF THE AM. METEOROLOGICAL SOC'Y (6), S19 ("Global average surface temperatures during the last three decades have been progressively warmer than all earlier decades, making 2000–09 (the 2000s) the warmest decade in the instrumental record."); L.V. Alexander et al., *Global Observed Changes in Daily Climate Extremes of Temperature and Precipitation*, J. GEOPHYSICAL RES., 111, D05109, doi:10.1029/2005JD006290, at 18 ("Between 1951 and 2003, over 70% of the land area sampled showed a significant increase in the annual occurrence of warm nights while the occurrence of cold nights showed a similar proportion of significant decrease. For the majority of the other temperature indices, over 20% of the land area sampled exhibits a statistically significant change....").

3. NATIONAL RESEARCH COUNCIL (NRC), SURFACE TEMPERATURE RECONSTRUCTIONS FOR THE LAST 2,000 YEARS (2006).

4. *Id.*

1988 by the United Nations Environment Programme (UNEP) and the World Meteorological Organization (WMO) and made up of thousands of scientists,[5] estimates that the average temperature will increase by 2 to 11.5 degrees Fahrenheit (1.1 to 6.4 degrees Celsius) by the end of the century.[6] The IPCC's best estimate range of projected increased temperatures was 3 to 7 degrees Fahrenheit (1.5 to 4 degrees Celsius).[7]

Climate change, however, does not just concern predicted higher atmospheric and oceanic temperatures. Climate change involves changes in temperature, precipitation, or wind patterns, among other effects, occurring over several decades or longer. While the majority of scientists agree that changes in temperatures at the Earth's surface are occurring, research related to changes in the frequency and severity of weather events linked to climate change are ongoing. Heat waves, tropical storms, hurricanes, tornadoes, and droughts are part of the natural cycle. The question is whether changes in the frequency and severity of these and other events are linked to climate change.

A recent report by scientists on the IPCC concluded the following regarding climate change:

- It is virtually certain that increases in the frequency and magnitude of warm daily temperature extremes and decreases in cold extremes will occur throughout the 21st century on a global scale.
- It is very likely that the length, frequency, and/or intensity of warm spells or heat waves will increase over most land areas.
- It is very likely that mean sea level rise will contribute to upward trends in extreme coastal high water levels in the future.
- Average tropical cyclone maximum wind speed is likely to increase, although increases may not occur in all ocean basins.
- It is likely that the frequency of heavy precipitation or the proportion of total rainfall from heavy falls will increase in the 21st century over many areas of the globe.[8]

---

5. The IPCC Fourth Assessment Report is just one of many sources that provides information on recent developments in the scientific understanding of climate change. Given this book's focus on insurance and the law, the voluminous peer-reviewed scientific literature on climate change is not detailed here. Other important reviews and sources of climate-change science include, but are not limited to, the National Climate Assessment (NCA) and the Bulletin of the American Meteorological Society's annual report, *State of the Climate*. The next update of the NCA is expected to be available in draft in late 2012 with final publication expected in 2013. The IPCC Fifth Assessment Report also likely will be published in 2013.

6. *See, e.g.*, Union of Concerned Scientists, Findings of the IPCC Fourth Assessment Report (2007), *available at* http://www.ucsusa.org/assets/documents/global_warming/IPCC-WGI-UCS-summary-300dpi.pdf.

7. *Id.*

8. IPCC, *2012: Summary for Policymakers*, Managing the Risks of Extreme Events and Disasters to Advance Climate Change Adaptation 11, 13 (C.B. Field, V. Barros, T.F. Stocker, D. Qin,

Scientists are observing an increase in the frequency and severity of weather-related events such as large rainstorms and heat waves. Attribution research continues to evaluate whether particular events are part of the natural and variable cycle or are tied to climate change.[9] Studies have found that the impacts of extreme weather events are also significantly confounded by economic and population growth, location choices (such as settlement in coastal zones and floodplains), and policy decisions that have led to more insured assets being placed in the paths of severe weather events.[10] Thus, the evolution of the scientific, demographic, and policy research is important in the insurance context.

B. The Importance of Definitions in the Insurance Context

Definitions are always important in an insurance context. Insurers and insureds want to achieve a mutual understanding regarding what is covered and what is not. As in any contract setting, issues arise when an insured and insurer have different understandings of the same term. The imprecise usage and varying definitions of the terms "climate change" and "global warming," therefore, could have implications in the insurance context.

Despite differences in the concepts, the terms "climate change" and "global warming" are often employed generically and interchangeably by consumers, the media, and policymakers. There is no single universally accepted definition of either term. Disparities in the use and meaning of such terms as well as conflicting estimations about the scope of changes and impacts cause a degree of legal uncertainty for insurers. "Global warming" is generally a reference to the trend of an increase in the average temperature near the Earth's surface.[11] Climate change is generally accepted to be something broader than global warming because it is not limited to changes in temperature.

D.J. Dokken, K.L. Ebi, M.D. Mastrandrea, K.J. Mach, G.K. Plattner, S.K. Allen, M. Tignor & P.M. Midgley eds., 2011), *available at* http://www.ipcc.ch/pdf/special-reports/srex/SREX_FD_SPM_final.pdf; *see also* Jane Lubchenco, Under Sec'y of Commerce for Oceans & Atmosphere and NOAA Adm'r, Predicting and Managing Extreme Events, at the American Geophysical Union (Dec. 7, 2011), *available at* http://www.noaanews.noaa.gov/stories2011/20111207_speech_agu.html.

9. *See, e.g.*, Quirin Schiermeier, *Climate and Weather: Extreme Measures: Can Violent Hurricanes, Floods and Droughts Be Pinned on Climate Change?*, 477 NATURE , 148–49 (2011); *see also* Malcolm Ritter, *Global Warming Tied to Extreme Weather Risk*, WALL ST. J., July 11, 2012 (discussing results of studies by researchers at Oregon State University and in England and by scientists from Oxford University and the British government assessing how climate change has changed the odds of extreme weather events happening).

10. Laurens Bouwer, *Have Disaster Losses Increased Due to Anthropogenic Climate Change?* BULL. OF THE AM. METEOROLOGICAL SOC'Y, Jan. 2011, at 39–46.

11. *See, e.g., Climate Change Basics*, EPA, http://www.epa.gov/climatechange/basics/ (last updated Aug. 21, 2012).

Some definitions involve an anthropogenic component. The IPCC, for example, has defined climate change as "a change in the state of the climate that can be identified (e.g., by using statistical tests) by changes in the mean and/or the variability of its properties, and that persists for an extended period, typically decades or longer. Climate change may be due to natural internal processes or external forcings, or to persistent anthropogenic changes in the composition of the atmosphere or in land use. . . ."[12]

To effectively address the risks associated with climate change, insurers and insureds will benefit from an awareness of the varying views and definitions of "climate change," "global warming," "greenhouse gases," and other related terms. A lack of understanding of the differences by insurers could lead to (a) underwriting mistakes; (b) difficulties in drafting forms; (c) claims relating to inconsistent positions based on varying statements of corporate, underwriting, marketing, claims, and loss-control consulting branches of a company; and (d) difficulties in internally communicating the extent of and solutions to the emerging risk. Definitions should be considered in the context of new insurance product development.

•••
Usage of different terminology and definitions concerning climate change and global warming create legal uncertainty.
•••

C. Greenhouse Gases and the Causation Issue

What is a GHG and how are GHGs related to the climate change issue? According to the IPCC, "GHGs are those gaseous constituents of the atmosphere, both natural and anthropogenic, that absorb and emit radiation at specific wavelengths within the spectrum of thermal infrared radiation emitted by the Earth's surface, the atmosphere itself, and by clouds."[13] Many chemical compounds found in the Earth's atmosphere act as GHGs. Some of them occur in nature (e.g., water vapor, carbon dioxide, methane, and nitrous oxide), while others are exclusively human-made (e.g., perfluorocarbons). The Earth's climate depends on the functioning of a natural "greenhouse effect."[14] Without this natural greenhouse effect, the average surface temperature of the Earth would be about 60 degrees Fahrenheit colder.[15] As explained by the National Oceanic and Atmospheric Administration (NOAA), the "greenhouse effect" is the "process which

---

12. IPCC, Climate Change 2007—The Physical Science Basis 943 (S. Solomon, D. Qin, M. Manning, Z. Chen, M. Marquis, K.B. Averyt, M. Tignor & H.L. Miller eds., 2007), *available at* http://www.ipcc.ch/pdf/assessment-report/ar4/wg1/ar4-wg1-annexes.pdf.

13. *Id.* at 947.

14. U.S. Global Change Research Program, Global Climate Change Impacts in the United States 14 (Thomas R. Karl, Jerry M. Melillo & Thomas C. Peterson eds., 2009), *available at* http://downloads.globalchange.gov/usimpacts/pdfs/climate-impacts-report.pdf.

15. *Id.*

warms the Earth's atmosphere due to the absorption of radiation energy by several trace gases; these greenhouse gases allow solar radiation to reach the Earth's surface but then absorb the energy as it is reemitted as infrared radiation, acting to contain the heat within the atmosphere."[16]

Scientists have concluded that "human activities have been releasing additional heat-trapping gases, intensifying the natural greenhouse effect, thereby changing the Earth's climate."[17] According to the IPCC, "[w]arming of the climate system is unequivocal, as is now evident from observations of increases in global average air and ocean temperatures, widespread melting of snow and ice, and rising global average sea level."[18] The IPCC contends that GHGs have very likely (more than 90 percent probability[19]) caused most of the observed global warming over the last half century.[20] The IPCC also concludes that there has probably (more than 66 percent probability) been a substantial anthropogenic or human contribution to surface temperature increases in every continent except Antarctica since the middle of the 20th century.[21] In other words and as discussed further below, organizations such as the IPCC and the National Academy of Sciences assert that recent climate change (within the past 50 years) has been, at least to some extent, caused by GHGs from human activity.

The scientific correlation of climate changes with increases in anthropogenic GHG emissions is sometimes referred to by scientists and organizations such as the IPCC as "attribution." Attribution more broadly refers to an analysis of what is causing changes. "Attribution" is defined by the IPCC as the process of establishing the most likely causes for detected[22] climate changes with some defined level of confidence.[23] The IPCC has examined "attribution" studies attempting to separate

---

16. National Oceanic & Atmospheric Administration, *Global Monitoring Division Education & Outreach*, http://www.esrl.noaa.gov/gmd/education/terms.html#G.

17. U.S. GLOBAL CHANGE RESEARCH PROGRAM, *supra* note 14.

18. IPCC, CLIMATE CHANGE 2007—SYNTHESIS REPORT 72 (R.K. Pachauri & A. Reisinger eds., 2007), available at http://www.ipcc.ch/pdf/assessment-report/ar4/syr/ar4_syr.pdf.

19. The IPCC uses standard terms in its report to define the likelihood of an outcome or result. The likelihood terminology and probability percentages are listed in Box 1.1 (Treatment of Uncertainties in the Working Group I Assessment) in IPCC, CLIMATE CHANGE 2007—THE PHYSICAL SCIENCE BASIS 120 (S. Solomon, D. Qin, M. Manning, Z. Chen, M. Marquis, K.B. Averyt, M. Tignor & H.L. Miller eds., 2007), available at http://www.ipcc.ch/pdf/assessment-report/ar4/wg1/ar4-wg1-chapter1.pdf.

20. *Id.* at 665, 728–29, available at http://www.ipcc.ch/pdf/assessment-report/ar4/wg1/ar4-wg1-chapter9.pdf.

21. *Id.*

22. The concept of "detection" is described by organizations, like the IPCC, as the process of demonstrating that the climate has changed in some defined statistical senses, without providing a reason for that change. According to the IPCC, an identified change is "detected" if the likelihood of its occurrence by chance is determined to be small. In other words, detection involves demonstrating that an observed change is statistically significantly different from that which can be explained by natural occurrence. IPCC, CLIMATE CHANGE 2007—THE PHYSICAL SCIENCE BASIS 668 (S. Solomon, D. Qin, M. Manning, Z. Chen, M. Marquis, K.B. Averyt, M. Tignor & H.L. Miller eds., 2007), available at http://www.ipcc.ch/pdf/assessment-report/ar4/wg1/ar4-wg1-chapter9.pdf.

23. *Id.*

natural variability from the impacts of anthropogenic emissions.[24] Scientists are working to create methodologies for attributing specific severe weather events to human-induced GHG emissions.[25]

In February 2011, for example, two studies were published in the journal *Nature* that suggest anthropogenic global warming is already playing a tangible role in influencing some types of extreme weather events. One study, led by researchers with Environment Canada, analyzed heavy rainfall events recorded across the Northern Hemisphere and found that human-induced GHG emissions have probably increased the intensity of heavy precipitation events.[26] The second study demonstrated a new way of analyzing how man-made global warming may have increased the chances for a particular flood that occurred in the United Kingdom in 2000.[27] In nine out of ten cases, the computer models indicate that 20th-century anthropogenic GHG emissions increased the risk of floods occurring in England and Wales in autumn 2000 by more than 20 percent, and in two out of three cases by more than 90 percent.[28]

Particularly in the United States, as compared with Europe or other parts of the world, there are some that question the science of climate change and the notion that the increase of GHG emissions from anthropogenic sources is causing global warming. Some scientists have developed new studies to try to address prior criticisms of the science.[29] Investigation of the exact causes of particular climate changes and the timing and scope of the impacts of increased GHG emissions continues.

D. Greenhouse Gases (GHGs)

For ease of regulation and management, most policymakers have chosen to focus on anthropogenic sources of seven categories of GHGs: carbon dioxide (CO_2), methane (CH_4), nitrous oxide (N_2O), perfluorocarbons (PFCs), sulfur hexafluoride (SF_6), hydrofluorocarbons (HFCs), as well as other fluorinated gases such as nitrogen trifluoride (NF_3) and fluorinated ethers. Some of the GHGs in the atmosphere contribute

---

24. *Id.* at 663–727.
25. *See, e.g.*, Peter A. Stott et al., *Attribution of Weather and Climate-Related Extreme Events* (World Climate Research Programme, Climate Change in Service to Society Conference, Position Paper, Oct. 27, 2011), *available at* http://conference2011.wcrp-climate.org/documents/Stott.pdf.
26. Seung-Ki Min, Xuebin Zhang, Francis W. Zwiers & Gabriele C. Hegerl, *Human Contribution to More-Intense Precipitation Extremes*, 470 Nature 378, Feb. 17, 2011.
27. Pardeep Pall et al., *Anthropogenic Greenhouse Gas Contribution to Flood Risk in England & Wales in Autumn 2000*, 470 Nature 382, Feb. 17, 2011.
28. *Id.*
29. For example, a team of University of California, Berkeley, physicists and statisticians undertook to review the temperature data underlying most global warming studies. In 2011, the Berkeley Earth Surface Temperature project released the group's findings, which tended to substantiate the findings of previous studies. *See* http://berkeleyearth.org (last visited January 30, 2012).

to the greenhouse effect more than others. Thus, scientists developed a way to weigh GHG emissions by their "global warming potential" so that contributions of various types of GHG emissions could be meaningfully compared.[30] According to the EPA Inventory of U.S. Greenhouse Gas Emissions and Sinks: 1990–2009, these first six categories of GHGs accounted in 2009 for the following percentage of U.S. emissions based on global warming potential: CO_2 (83 percent); CH_4 (10.3 percent); N_2O (4.5 percent); and PFCs, SF_6, and HFCs (collectively 2.2 percent).[31]

It is important to note for risk management and underwriting purposes, however, that many other gases, such as water vapor (H_2O), ozone (O_3), and black carbon, can act as GHGs.[32] Thus, when designing GHG exclusions or GHG coverage, especially definitions of GHGs or climate change, it is important to consider that a definition in an exclusion restricted to the main categories of GHGs may not be comprehensive and may leave unmanaged or unaddressed exposures. That said, a broad exclusion may also unintentionally eliminate core coverage related to standard coverage items such as for fire or water damage not otherwise excluded through a mold and fungi exclusion. By contrast, a broad definition set forth in a policy providing affirmative coverage for GHG-related issues may provide coverage for both regulated and unregulated GHGs and GHGs from both natural and anthropogenic sources.

The sources of seven of the most targeted GHGs (for regulatory purposes) as well as water vapor are:

(1) Carbon dioxide (CO_2). When released into the atmosphere, CO_2 forms a barrier, which traps heat that would otherwise escape into space. The largest source of anthropogenic CO_2 is the burning of fossil fuels (coal, oil, and natural gas).[33] Most of the burning occurs in automobiles, power

---

30. IPCC has defined "global warming potential" or "GWP" as an "an index, based upon radiative properties of well-mixed greenhouse gases, measuring the radiative forcing of a unit mass of a given well-mixed greenhouse gas in the present-day atmosphere integrated over a chosen time horizon, relative to that of carbon dioxide. The GWP represents the combined effect of the differing times these gases remain in the atmosphere and their relative effectiveness in absorbing outgoing thermal infrared radiation." IPCC, *Climate Change 2007—The Physical Science Basis* 946 (S. Solomon, D. Qin, M. Manning, Z. Chen, M. Marquis, K.B. Averyt, M. Tignor & H.L. Miller eds., 2007), *available at* http://www.ipcc.ch/pdf/assessment-report/ar4/wg1/ar4-wg1-annexes.pdf.

31. EPA, Inventory of U.S. Greenhouse Gas Emissions and Sinks: 1990–2009, Executive Summary at 7, fig.ES-4 (Apr. 2011).

32. IPCC, Climate Change 2007—The Physical Science Basis 33–34 (S. Solomon, D. Qin, M. Manning, Z. Chen, M. Marquis, K.B. Averyt, M. Tignor & H.L. Miller eds., 2007), *available at* http://www.ipcc.ch/pdf/assessment-report/ar4/wg1/ar4-wg1-ts.pdf. *See also* Chapter 4.B.1 and 5.B.4, discussing EPA's decision not to include certain GHGs in its GHG Mandatory Reporting Rule.

33. EPA, *Carbon Dioxide Emissions*, http://www.epa.gov/climatechange/ghgemissions/gases/co2.html.

plants, and industrial factories such as steel, iron, cement, and refining.[34] Land-use changes, such as from deforestation, also have resulted in significant emissions of CO_2.[35] A very small amount of CO_2 is also naturally emitted by humans and animals.

(2) Methane (CH_4). CH_4 is emitted from a variety of both anthropogenic and natural sources. Anthropogenic sources include fossil-fuel production such as coal mining, landfills, biomass burning, and ruminant digestion and manure management associated with domestic livestock operations.[36] These sources release significant quantities of CH_4 to the atmosphere. According to some estimates, at least 50 percent of global CH_4 emissions are related to human activities.[37]

(3) Nitrous oxide (N_2O). N_2O is produced by both natural and human-related sources. Primary human-related sources of N_2O include agricultural soil management, animal manure, sewage treatment, combustion of fossil fuel, adipic acid production, and nitric acid production.[38] N_2O emission levels from a source can vary significantly from one country or region to another, depending on fertilizer usage, combustion technologies, waste management practices, and climate.[39] The two main natural sources of N_2O emissions come from the bacterial breakdown of nitrogen in soil and in the oceans.[40]

(4) Perfluorocarbons (PFCs). All PFCs are man-made. Primary aluminum production and semiconductor manufacturing are the largest known man-made sources of two PFCs—tetrafluoromethane (CF_4) and hexafluoroethane (C_2F_6).[41]

(5) Sulfur Hexafluoride (SF_6). SF_6 has a high GWP.[42] SF_6 is used (1) for insulation and current interruption in electric power transmission,

---

34. *See, e.g.*, EPA, *Quantifying Greenhouse Gas Emissions from Key Industrial Sectors in the United States*, May 2008, *available at* http://www.epa.gov/sectors/pdf/greenhouse-report.pdf.

35. *See, e.g.*, G.J. Nabuurs, O. Masera, K. Andrasko, P. Benitez-Ponce, R. Boer, M. Dutschke, E. Elsiddig, J. Ford-Robertson, P. Frumhoff, T. Karjalainen, O. Krankina, W.A. Kurz, M. Matsumoto, W. Oyhantcabal, N.H. Ravindranath, M.J. Sanz Sanchez & X. Zhang, 2007: Forestry in Climate Change 2007: Mitigation. Contribution of Working Group III to the Fourth Assessment Report of the Intergovernmental Panel on Climate Change (B. Metz, O.R. Davidson, P.R. Bosch, R. Dave & L.A. Meyer eds., Cambridge Univ. Press, 2007).

36. EPA, Sources and Emissions, http://www.epa.gov/methane/sources.html (last updated April 18, 2011).

37. *Id.*

38. EPA, Sources and Emissions, http://www.epa.gov/nitrousoxide/sources.html (last updated June 22, 2010).

39. *Id.*

40. *Id.*

41. EPA, Science: High GWP Gases and Climate Change, http://www.epa.gov/highgwp1/scientific.html (last updated Feb. 9, 2011).

42. *Id.*

(2) in distribution equipment in the magnesium industry to protect molten magnesium from oxidation and potentially violent burning, (3) in semiconductor manufacturing to create circuitry patterns on silicon wafers, and (4) as a tracer gas for leak detection.[43]

(6) Hydrofluorocarbons (HFCs). HFCs are produced primarily through chemical manufacturing, construction, and automotive air conditioning. HFCs, which do not contain ozone-destroying chlorine or bromine atoms, are used as substitutes for ozone-depleting compounds such as chlorofluorocarbons (CFCs) for refrigeration, air conditioning, and the production of insulating foams. The 1987 Montreal Protocol has gradually phased out the use of CFCs and other ozone-depleting substances, leading to the development and increased use of long-term replacements such as HFCs.[44]

(7) Other fluorinated gases such as nitrogen trifluoride (NF_3) and fluorinated ethers (HFEs and HCFEs). NF_3 is used in the semiconductor industry and to produce LCD flat-screen TVs and solar photovoltaic cells. HFEs and HCFEs are used as anesthetics (e.g., isoflurane, desflurane, and sevoflurane) and as heat transfer fluids (e.g., the H-Galdens).[45]

(8) Water vapor (H_2O). Water vapor is a significant GHG, accounting for approximately 60 percent of the greenhouse effect.[46] As the temperature of the atmosphere rises, more water is evaporated from ground storage (rivers, oceans, reservoirs, soil). Warmer air has a higher capacity to retain water vapor.[47] As a GHG, the higher concentration of water vapor is then able to absorb more thermal energy, thus further heating the atmosphere. The warmer atmosphere can then hold even

---

43. MINN. POLLUTION CONTROL AGENCY, TECHNICAL EVALUATION OF THE EMISSIONS AND CONTROL COSTS OF HIGH GLOBAL WARMING POTENTIAL GASES 11 (Feb. 1, 2009), *available at* www.leg.state.mn.us/docs/2009/other/090493.pdf.

44. The Montreal Protocol on Substances That Deplete the Ozone Layer (a protocol to the Vienna Convention for the Protection of the Ozone Layer) is an international treaty designed to protect the ozone layer by phasing out the production of a number of substances believed to be responsible for ozone depletion. The treaty entered into force on Jan. 1, 1989.

45. EPA, Technical Support Document for Industrial Gas Supply: Production, Transformation, and Destruction of Fluorinated GHGs and N_2O Proposed Rule for Mandatory Reporting of Greenhouse Gases at 3 (Feb. 6, 2009), *available at* http://nepis.epa.gov/EPA/html/DLwait.htm?url=/Exe/ZyPDFcgi?Dockey=P1009B30.PDF.

46. IPCC, CLIMATE CHANGE 2007—THE PHYSICAL SCIENCE BASIS 271 (S. Solomon, D. Qin, M. Manning, Z. Chen, M. Marquis, K.B. Averyt, M. Tignor & H.L. Miller eds., 2007), *available at* http://www.ipcc.ch/pdf/assessment-report/ar4/wg1/ar4-wg1-chapter3.pdf. Water vapor, which is naturally occurring, is not currently regulated as a GHG.

47. THE INSTITUTE FOR CATASTROPHIC LOSS REDUCTION, TELLING THE WEATHER STORY 10 (Insurance Bureau of Canada, June 2012).

more water vapor. This is referred to as a "positive feedback loop."[48] The Energy Information Administration (EIA) asserts that although water vapor is the most abundant GHG, human influence on water vapor's concentration in the atmosphere is thought by scientists to be negligible.[49]

E. Conclusion

Both the science related to the climate change issue and the nature and degree of regulatory focus upon various GHGs continue to evolve. Insurers and insureds seeking to assess the potential impact of climate change on their respective industries would do well to keep abreast of new developments in these areas. This is true regardless of whether scientists and other observers reach consensus regarding the issue of climate change in general or anthropogenic causation specifically. As explored in the ensuing chapters, the emerging risk of climate change is not dependent on achieving scientific or regulatory certainty.

---

48. Scientific uncertainty exists in defining the extent and importance of the water vapor feedback loop. As water vapor increases in the atmosphere, more of it will eventually also condense into clouds, which are more able to reflect incoming solar radiation, thus allowing less energy to reach the Earth's surface.

49. Energy Information Administration, Emissions of Greenhouse Gases in the United States (2007) (released Dec. 3, 2008).

CHAPTER 3

Climate Change as an Emerging Risk and Comparison to Historic Emerging Risks

A. Climate Change as an Emerging Risk

An emerging risk is one that has not yet fully developed and is not yet conducive to measurement. Swiss Re, for example, defines emerging risks as "newly developing or changing risks which are difficult to quantify and which may have a major impact on an organisation."[1] Lloyd's defines an emerging risk as "an issue that is perceived to be potentially significant but which may not be fully understood or allowed for in insurance terms and conditions, pricing, reserving or capital setting."[2]

Many refer to climate change as an "emerging risk" because there is uncertainty with respect to the parameters of the issue from a scientific, political, social, and economic perspective. Stakeholders, such as GHG emitters and insurers, for example, do not fully know what legal regimes may be created to achieve adaptation to and mitigation of climate change-related risks. As with all emerging issues, affected parties cannot definitively estimate future costs when there is no extensive history of regulation or claims.

At the time insurers underwrote many policies currently in force, including historic-occurrence policies, insurers may not have been aware of all the potential impacts of the climate change-related issue on the insurance industry. Thus, climate change risk may not have been contemplated in developing policy language, underwriting risk, or developing pricing models. The potential exposures associated with climate-change impacts, however, are now beginning to emerge. While uncertainty in the marketplace exists, it is clear that over the coming years the insurance market will be affected by climate change in some way, shape, or form.

---

1. *Identifying Emerging Risks*, Swiss Re, http://www.swissre.com/rethinking/emerging_risks/ (last visited Feb. 1, 2012).
2. *Emerging Risks*, Lloyd's, http://www.lloyds.com/The-Market/Tools-and-Resources/Research/Exposure-Management/Emerging-risks (last visited Feb. 1, 2012).

B. Evolution of Other Emerging Risks

There are historical examples of risks that initially did not gain traction in the courts or political forums but eventually resulted in significant recoveries. Three well-known and well-documented patterns of litigation (asbestos, tobacco, and CERCLA/Superfund litigation) may inform an evaluation of the emerging climate change-related liability risks. In attempting to develop successful lawsuits, the plaintiffs' bar also will look to past successful strategies on emerging issues. Gerald Maples, the lead plaintiffs' attorney in *Comer v. Murphy Oil*, one of the first major climate change-related tort cases, stated that he based his lawsuit against insurers, large chemical companies, and oil companies on tobacco litigation: "What's good about the approach that I'm taking is that the tobacco litigation—and before that the asbestos litigation—demonstrates that one case can cause a gigantic litigation problem for corporations. It's pretty much accepted history that asbestos and tobacco are the role models for climate change litigation now."[3] Mr. Maples' comment reveals the plaintiffs' bar's awareness of comparing past emerging risks. The emerging tort litigation is discussed in more detail in Chapter 6.A.

Factual differences between tobacco, asbestos, and CERCLA/Superfund claims and climate change may result in different drivers and patterns of claims. At a minimum, the vast number and variety of economic sectors and demographics affected by climate change are far greater than in any of these prior risk experiences. Nevertheless, it is worth considering whether history could repeat itself with respect to the development of mega-claims.

In 1987, Bruce A. Levin published an article in the *Arizona Law Review* that stated: "Tobacco companies boast that they have never lost a case to a consumer, have never settled, and do not expect that picture to change. In the 1950s and 1960s, no cases successfully obtained damages for injuries caused by smoking. Recent cases have been similarly unsuccessful."[4]

Of course, as predicted by Mr. Levin, things changed. History reflects that for many years, the tobacco industry insisted that the causal connection between a specific pack (or packs) of cigarettes and an individual's cancer was too tenuous to meet the burden for recovery in a tort action. Even as "the abundance of materials demonstrating the hazards of smoking" mounted, "tobacco companies . . . steadfastly maintained

---

3. Paddy Manning, *You Are at Risk*, THE SYDNEY MORNING HERALD (June 20, 2009), http://www.smh.com.au/business/you-are-at-risk-20090620-crk4.html.

4. Bruce A. Levin, *The Liability of Tobacco Companies—Should Their Ashes Be Kicked?*, 29 ARIZ. L. REV. 195, 200 (1987).

that unbiased research [was] needed to resolve the health 'controversy.'"[5] Indeed, Mr. Levin noted in 1987: "Despite an almost endless supply of evidence documenting the hazards of smoking, tobacco companies continue to deny that their products are harmful."[6]

If past is prologue, a similar story could unfold with respect to emerging climate change litigation and the position of GHG emitters. Industry has already experienced a U.S. Supreme Court victory on one claim.[7] Ultimately, the tobacco industry succumbed to the mountain of litigation with the payment of billions of dollars in a mega-settlement of claims extending from states to individuals in massive class actions. Claims continue to generate huge verdicts.

As with tobacco, the combination of a growing consensus among scientists, together with administrative governmental determinations in the face of deliberate congressional inaction, could create the foundation for mega-recoveries in the climate change arena. For tobacco, the mega-recoveries followed years of near uniformity among researchers (over continuous challenges of bias), culminating in determinations by the surgeon general that tobacco indeed contributed to cause an increased risk of cancer.

Importantly, neither the surgeon general nor any court found that tobacco was "the" cause of cancer. Instead, as Mr. Levin noted, "[t]he Surgeon General's reports, both 1964 and 1979, conclude[d] that cigarette smoking is a cause, not the cause."[8] Similarly, the center of gravity in the climate change debate is now whether GHG emissions are "a" cause of climate change as opposed to "the" cause.

Increasingly, climate change claimants (and the plaintiffs' attorney bar) will undoubtedly attempt to reframe the climate change issue in the context of collective redress, including judicial remedies, as one simple question: Should those who have profited the most from the release of GHG emissions have to share some of their gains with those who have suffered the most?

Within the context of this justification for recovery, evolution of climate litigation might follow a pattern similar to historical mega-risks. Typically, mega-exposures like tobacco, asbestos, and pollution follow a predictable path evolving from isolated, untested claims to huge payments on a class or national basis.

Based on these historical patterns, the five phases of mass tort recoveries could be described as:

---

5. *Id.* at 198.
6. *Id.* at 195.
7. Am. Elec. Power Co. v. Connecticut, 131 S. Ct. 2527 (2011).
8. Levin, *supra* note 4, at 223, n.223.

Phase I: Prospecting—Unsuccessful, intermittent strike claims based on myriad traditional tort recovery theories, designed largely to explore the boundaries for successful recoveries.

Phase II: Defining—Increased regulatory activity supplying standards by which the standard of care and causation can be established, accompanied by increasing numbers of adapted claims.

Phase III: Refining—More sophisticated complaints supported by well-funded plaintiffs' attorneys, causing increased discovery costs and resulting in occasional rulings that permit claims to reach finders of fact.

Phase IV: Targeting—Intermittent settlements as litigation costs begin to systematically exceed discovery costs and vulnerable, targeted defendants are found and fall.

Phase V: Recovering—Plaintiffs' attorneys accumulate enough resources and data to evenly battle industry targets, culminating in the ultimate collapse of industry targets.

Like similar mega-exposures, successful climate change recoveries will not occur overnight. Early climate change claimants face seemingly insurmountable odds in their bid to recover from GHG emitters. They face numerous legal hurdles including the political question doctrine defense, standing challenges, difficulties in proving causation, and displacement or preemption of claims by federal laws and regulations. With estimated damages in the trillions of dollars, claimants or plaintiffs' attorneys likely will not simply retreat.[9] When one theory does not succeed, other theories of recovery will follow. Claims may persist, even as federal common law of nuisance or even state law nuisance claims are determined not to be viable. For example, just as the Supreme Court was denying a federal common law nuisance claim, new plaintiffs were filing cases based on the public trust doctrine.[10]

A worrisome trend for GHG emitters and insurers has emerged. Insurance industry representatives and regulators are beginning to take note of the potential risks associated with emerging climate change litigation. GHG emitters are considering what path to take in the midst or in the absence of regulation. The question now is whether and how history will repeat itself.

---

9. The United Nations Environmental Programme Finance Initiative (UNEP FI) and Principles for Responsible Investment (PRI) issued a report finding that global environmental damage caused by human activity in 2008 (including from climate change) represented a monetary value of $6.6 trillion and that the world's top 3,000 public companies were responsible for a third of all global environmental damage. *See* UNEP FI & PRI, Universal Ownership: Why Environmental Externalities Matter to Institutional Investors (2010), *available at* http://www.unepfi.org/fileadmin/documents/universal_ownership.pdf.

10. *See* Chapter 6.C.

CHAPTER 4

The Likely Plaintiffs and Targets of Emerging Climate Change-Related Litigation and Regulation

A. Potential Impacts and Likely Claimants

As climate change-related litigation heats up, potential claims are most likely to emerge from the following categories of individuals or entities: (1) persons suffering property damage, bodily injury, or other damages as a result of episodic climatic events such as hurricanes, droughts, and fires; (2) residents of low-lying and coastal regions; (3) NGOs; (4) U.S. states, cities, and other governmental entities; (5) industries particularly affected by climate change such as tourism, skiing, agriculture, forestry, and fishing; (6) shareholders and investors; and (7) other insurers (for subrogation).

Indeed, representatives of several of the above categories already have pursued or are pursuing climate change-related litigation. As discussed in more detail in Chapter 6, several U.S. states, cities, and other governmental entities brought suit seeking to curtail GHG emissions from utilities across state borders (*American Electric Power Co. v. Connecticut*[1]); residents of a coastal Alaskan village are seeking monetary damages and other relief from the energy industry for the destruction of their ancestral homeland due to flooding allegedly caused by climate change (*Kivalina v. ExxonMobil Corp.*[2]); and a group of Gulf Coast property owners filed a lawsuit against energy companies for their contributions to climate change, which the plaintiffs claimed contributed to the severity of Hurricane Katrina (*Comer v. Murphy Oil USA, Inc.*[3]). NGOs also have begun to enter the fray, as evidenced by a public interest group's filing

---

1. See Chapter 6.A.2 (discussing *Am. Elect. Power Co.*). See Am. Elec. Power Co. v. Connecticut, 131 S. Ct. 2527 (2011).
2. Native Vill. of Kivalina v. ExxonMobil Corp., No. 09-17490, ___ F.3d ___, 2012 WL 4215921 (9th Cir. Sept. 21, 2012), *petition for rehearing* filed Oct. 4, 2012. See Chapter 6.A.3 (discussing *Kivalina*).
3. Comer v. Nationwide Mut. Ins. Co., No. 1:05 CV 436 LTD RHW, 2006 WL 1066645, at *1 (S.D. Miss. Feb. 23, 2006), *rev'd sub nom.* Comer v. Murphy Oil USA, 585 F.3d 855 (5th Cir. 2009), *rev'd en banc*, 607 F.3d 1049 (5th Cir. 2010). See Chapter 6.A.5.b (discussing *Comer*).

in 2011 of suits or petitions in all 50 states seeking to reinvigorate federal and state regulation to combat climate change.[4]

The plaintiffs in the *American Electric Power Co. v. Connecticut* case identified a number of future injuries associated with climate change that are illustrative of the types of claims that might precipitate future litigation in this area. With regard to personal injuries, for example, the plaintiffs alleged that more intense and/or prolonged heat waves could result in increased illnesses and deaths, and increased smog could increase or exacerbate residents' respiratory problems.[5] In terms of property damage and/or business-related claims, the plaintiffs pointed to (among other things) damage to coastal property and infrastructure due to significant beach erosion and/or accelerated rise of sea level; impaired shipping, recreational use, and hydropower generation due to lowered Great Lakes water levels; and widespread damage to property and ecosystems due to increased wildfires.[6]

B. Targets/Defendants

1. Emitters of GHGs

Industries emitting the largest amounts of GHGs are the most likely targets of climate change-related regulation and climate change-related litigation claims. According to EPA, the majority of GHG emissions in the United States come from the sectors shown in Figure 1.

As evidenced by the kinds of climate change-related lawsuits brought to date (see Chapter 6), electric power generators and other industrial concerns already have emerged as key targets. EPA's GHG Reporting Rule,[7] which requires thousands of affected facilities to submit certified annual reports detailing emissions of CO_2, CH_4, and N_2O; fluorinated gases such as SF_6, HFCs, PFCs, and NF_3; and hydrofluorinated ethers (HFEs), provides further information regarding which industries likely will be the subject of future GHG regulation and litigation. Although the rule itself does not require control of GHG emissions and demands only that GHG sources monitor and report emissions above certain threshold levels, the EPA reporting system makes GHG emissions data readily available for regulators, future plaintiffs, and the investment community.[8]

---

4. The legal actions coordinated by Our Children's Trust, an Oregon-based public interest group, are grounded in the Public Trust Doctrine, which holds that it is the government's duty to protect the resources that are essential for collective survival and prosperity. *See* http://ourchildrenstrust.org/legal-action (last visited January 31, 2012). *See also* Chapter 6.C.

5. Connecticut v. Am. Elec. Power Co., 582 F.3d 309, 318 (2d Cir. 2009).

6. *Id.*

7. The final rule for some source categories was signed by Administrator Jackson on Sept. 22, 2009, and became effective on Dec. 29, 2009. *See* Mandatory Reporting of Greenhouse Gases, 74 Fed. Reg. 56,260, 56,267 (Oct. 30, 2009) (codified at 40 C.F.R. pts. 86, 87, 89, et al.).

8. *2010 GHG Emissions from Large Facilities*, EPA, http://ghgdata.epa.gov/ghgp/main.do.

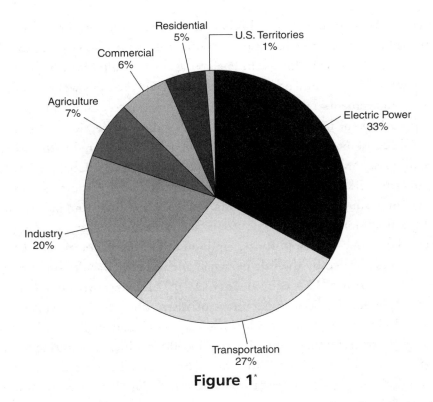

Figure 1*

---

* *See* EPA, INVENTORY OF U.S. GREENHOUSE GAS EMISSIONS AND SINKS: 1990–2009, Executive Summary at 18, fig. ES-13 (Apr. 2011). Due to rounding, note that the EPA figures for each sector add up to 99 percent rather than 100 percent.

Generally, the EPA GHG Reporting Rule, as initially crafted, applies to fossil fuel and industrial GHG suppliers (upstream sources), direct emitters of GHGs (downstream sources), and manufacturers of certain mobile sources and engines. The initial reports under the rule covered calendar year 2010.[9] For some source categories added by a later amendment to the rule, monitoring requirements began in 2011 and initial reports were due on March 31, 2012.[10] Vehicle and engine manufacturer (such as heavy-duty trucks, motorcycles, and non-road engines) reporting began with model year 2011. Generally, sources must report at the facility level. That said, reporting at the corporate level will be required for certain supplier source categories, including

---

9. Final Regulation Extending the Reporting Deadline for Year 2010 Data Elements Required Under the Mandatory Reporting of Greenhouse Gases Rule, 76 Fed. Reg. 14,812 (March 18, 2011) (codified at 40 C.F.R. pt. 98).

10. Mandatory Reporting of Greenhouse Gases: Petroleum and Natural Gas, 75 Fed. Reg. 74,458 (Nov. 10, 2010) (codified at 40 C.F.R. pt. 98); Mandatory Reporting of Greenhouse Gases: Carbon Dioxide Injection and Geologic Sequestration, 75 Fed. Reg. 75,060 (Dec. 1, 2010) (codified at 40 C.F.R. pts. 72, 78, 98); Mandatory Reporting of Greenhouse Gases: Additional Sources of Fluorinated GHGs, 75 Fed. Reg. 74,774 (Dec. 1, 2010) (codified at 40 C.F.R. pt. 98).

importers of fuels or industrial GHGs and manufacturers of affected vehicles and engines.[11] Through an amendment to the initial rule, EPA also requires that facilities report the name, physical address, and percentage ownership of the highest-level parent companies.[12] Ceres[13] and other investor NGOs will probably work to link the EPA reporting requirements with SEC and other reporting requirements, which could have implications for D&O liability insurance. Additional details on the standard of care for GHG monitoring and the potential penalties associated with this reporting rule are discussed in the regulatory developments chapter. See Chapter 5.B.4.

Upstream users impacted by this rule include producers, importers, and exporters of petroleum products, natural gas liquids, coal-to-liquid fuels, and industrial GHGs; natural gas fractionators and distributors; facilities capturing or extracting CO_2 for commercial or sequestration purposes; and importers and exporters of CO_2.[14] Direct GHG emitters affected by the rule primarily include large facilities emitting at least 25,000 tons per year of CO_2 equivalent (CO_2e).[15] Mobile sources required to report under the rule consist of manufacturers of heavy-duty and off-road vehicles and engines, including heavy trucks, motorcycles, and non-road engines and equipment.[16] In total, EPA estimates that approximately 13,000 facilities, accounting for about 85 to 90 percent of GHGs emitted in the United States, will be covered under this rule.[17]

Examples of industries subject to the initial EPA GHG Reporting Rule are summarized in the table below.

---

11. Mandatory Greenhouse Gas Reporting Rule, 74 Fed. Reg. at 56,264.

12. Mandatory Reporting of Greenhouse Gases, 75 Fed. Reg. 57,669, 57,685 (Sept. 22, 2010) (codified at 40 C.F.R. pts. 98.2, 98.3, and 98.6).

13. *See* Chapter 6.E.1 for a detailed discussion about Ceres.

14. Mandatory Reporting of Greenhouse Gases, 74 Fed. Reg. at 56,267.

15. Pursuant to EPA's climate change glossary of terms, *available at* http://www.epa.gov/climatechange/glossary.html, a CO_2 equivalent (CO_2e) is a "metric measure used to compare the emissions from various greenhouse gases based upon their global warming potential." Weight would be an inadequate method of comparison because of varying GWPs of these cases. Under this system, one metric ton of CO_2e is considered to be the amount of a given greenhouse gas that makes the same contribution to global warming as a metric ton of CO_2. For example, one metric ton of CH_4 generally is recognized to equal 25 tons of CO_2e. The EPA rule does not specifically contain a 25,000 metric ton of CO_2e volume requirement for the following "all-in" source categories: adipic acid production, aluminum production, ammonia manufacturing, cement production, electricity generation (facilities that otherwise report CO_2 emissions year round), HCFC-22 production, HFC-23 destruction, lime manufacturing, nitric acid production, petrochemical production, petroleum refineries, phosphoric acid production, silicon carbide production, soda ash production, and titanium dioxide production. *See* Mandatory Reporting of Greenhouse Gases, 74 Fed. Reg. at 56,266–56,267. EPA justified this decision by noting that nearly all facilities with these source categories will emit greater than 25,000 metric tons of CO_2e per year. *Id.* at 56,272 n.16.

16. *Id.* at 56,352. Light-duty vehicles, including light-duty trucks and medium-level passenger vehicles, are not included in the GHG Reporting Rule. EPA, however, issued a rule applicable to light-duty vehicles. *See* Chapter 5.B.2.

17. *Fact Sheet, Mandatory Reporting of Greenhouse Gases*, EPA 1 (June 2011), http://www.epa.gov/climatechange/emissions/downloads09/FactSheet.pdf.

Upstream Sources	Producers; importers and exporters of coal-based liquid fuels
	Natural gas fractionators and local natural gas distribution companies
	Producers; importers and exporters of petroleum products
	Producers; importers and exporters of industrial GHGs (N_2O, CO_2)
Downstream Sources	General stationary fuel combustion sources (boilers, stationary internal-combustion engines, process heaters, combustion turbines, other stationary fuel-combustion equipment)
	Electricity generation
	Adipic acid production
	Aluminum production
	Ammonia manufacturing
	Cement production
	Ferroalloy production
	Glass production
	HCFC-22 production and HFC-23 destruction
	Hydrogen production
	Iron and steel production
	Lead production
	Lime manufacturing
	Miscellaneous uses of carbonates
	Nitric acid production
	Petrochemical production
	Petroleum refineries
	Phosphoric acid production
	Pulp and paper manufacturing
	Silicon carbide production
	Soda ash manufacturing
	Titanium dioxide production
	Zinc production
	Municipal solid waste landfills
	Manure management*
Mobile Sources	Vehicles and engines outside of light-duty sector
	Heavy-duty, non-road, aircraft, locomotive, and marine diesel engine manufacturing
	Heavy-duty vehicle manufacturing
	Small non-road and marine spark-ignition engine manufacturing
	Personal watercraft manufacturing
	Motorcycle manufacturing

*The agriculture sector, with the exception of the approximately 100 manure-management operations at livestock operations with GHG emissions that meet or exceed the threshold of 25,000 metric tons of CO_2e, are not covered by the rule. *See id.* at 2.

By amendment to the initial rule, EPA requires additional categories of sources to report their GHG emissions. For example, as of January 1, 2011, the following industries had to begin reporting their GHG emissions: magnesium production, underground coal mines, industrial landfills, wastewater treatment, oil and natural gas systems,[18] electronics manufacturing, fluorinated gas production, electrical transmission and distribution equipment use, importers and exporters of fluorinated GHGs inside precharged equipment or closed-cell foams, and electrical equipment manufacture or refurbishment.[19] Plaintiffs may target these industries as well. Although EPA considered including additional types of facilities in the program, EPA has declined to include the ethanol industry, suppliers of coal, or the food-processing industry as specific subparts within the GHG Reporting Rule. Nonetheless, plaintiffs could target these industries as well.[20]

2. Service Providers to Targeted Emitters

Consultants, architects, engineers, accountants, contractors, developers, alternative energy developers and providers, and commodity traders and brokers ("service providers") will make up a new target class for potential climate change-related litigation. As federal and state entities require measurement of GHG emissions, reporting of GHG emissions, and plans to mitigate GHG emissions, service providers will be called on to prepare GHG inventory reports and mitigation strategies. Companies rebuilding after episodic climatic events may call on service providers to replace previous structures with "green" buildings meeting green building standards such as the U.S. Green Building Council's (USGBC) LEED rating system. Companies offering low-emissions solutions such as solar, wind, and operation of facilities for high-emitting power-generation plants also will continue to crop up. Trading and brokerage services for handling the transfer of carbon credits also will develop as international, federal, and state trading schemes get off the ground. If service providers make mistakes or

---

18. Oil and natural gas systems include offshore and onshore petroleum and natural gas producers, onshore natural gas processors, onshore natural gas transmission compressors, underground natural gas storage, liquefied natural gas storage, and liquefied natural gas import on export operations. See Mandatory Reporting of Greenhouse Gases: Petroleum and Natural Gas Systems, 75 Fed. Reg. 74,458 (Nov. 30, 2010).

19. See id.; see also Mandatory Reporting of Greenhouse Gases from Magnesium Production, Underground Coal Mines, Industrial Wastewater Treatment, and Industrial Waste Landfills, 75 Fed. Reg. 39,736 (July 12, 2010); Mandatory Reporting of Greenhouse Gases: Additional Sources of Fluorinated GHGs, 75 Fed. Reg. 74,774 (Dec. 1, 2010).

20. Even facilities that are not specifically included in a subpart of the GHG Reporting Rule are required to report if their aggregate emissions from listed source categories exceed the threshold. For example, a producer may be required to report if it emits over 25,000 metric tons of CO_2 after aggregating its emissions from the applicable source categories, such as industrial waste landfills and stationary combustion. *Frequently Asked Questions: Mandatory Reporting of Greenhouse Gases*, EPA 1 (June 2010), http://www.epa.gov/climatechange/emissions/downloads10/RS_FAQ.pdf.

their proposed solutions do not work or result in injury, lawsuits and insurance claims are likely to follow.

3. Impact to Insurers

Insurers may face two kinds of liability arising out of the climate change issue: (1) potential exposures arising out of institutional statements and activities ("enterprise liability"), and (2) potential exposures arising out of relationships with insureds ("potential claims liability"). The first potential core exposure arises from corporate functions such as climate change disclosures and development of knowledge of climate change-related issues or the provision of climate change-related consulting services. The second potential core exposure develops when insureds faced with climate change-related claims look to their existing insurance policies for coverage. If plaintiffs raise the myriad expected climate change-related claims, described in more detail in Chapter 6, then defendants certainly will look to their insurers for indemnity and defense coverage pursuant to existing policies and new products.

CHAPTER 5
Existing and Emerging Climate Change-Related Regulation

Over the course of the past two decades, regulation of GHGs has begun at the international, federal, regional, state, and local levels. Recently, however, efforts to regulate GHGs have increased, particularly in 2009. The evolving regulation of GHGs could be relevant to insurers for several reasons. First, as noted previously, regulation of GHG emissions will identify the key players in climate change-related disputes. Even if responses to regulatory mandates are determined to be clearly excluded under existing policies, increasing GHG regulation will make certain identified industries appetizing targets for the plaintiffs' bar. Insurers should identify and continue to follow which of their classes of insureds are subject to GHG regulation, because they may want to consider those facts in the underwriting process. Second, climate change-related environmental regulation will define terms and set standards that also might be relevant in the insurance context. For example, regulatory treatment of certain GHGs as "pollutants" could have significant impacts on the interpretation of pollution exclusions and affirmative grants of coverage for pollution events. Third, regulatory standards are relevant to insurers because they may be adopted as standards of care or used in causation analyses in the emerging climate change-related tort litigation. Whether climate change-related tort litigation succeeds or fails will depend in part on how courts use the evolving regulatory standards. Thus, insurers would be well advised to watch the regulatory developments, as they may indirectly affect the climate change-related litigation that could result in more indemnity and defense claims. Fourth, risks and opportunities may arise out of carbon trading schemes at the international, national, regional, and state levels. Finally, state insurance commissioners have become involved in climate change issues, particularly with respect to disclosure of risk.

Below are summaries of the recent major international, federal, regional, state, and local climate change-related regulatory events and developments.

A. The International Backdrop

Although many nations have not yet agreed to international binding emissions-reduction targets, international activities on climate change are significant and likely to play a greater role in years to come. Most international activities on climate change arise out of the United Nations Framework Convention on Climate Change (UNFCCC), an international environmental treaty signed in June 1992. The goal of the UNFCCC is to stabilize GHG concentrations in the atmosphere at a level that would prevent dangerous anthropogenic interference with the climate system.[1] The treaty as originally framed, however, set no mandatory limits on GHG emissions for individual nations and contained no enforcement provisions. Rather, the treaty included provisions for updates (called "protocols") that would set mandatory emission limits in the future and as appropriate. Guiding principles focus on a precautionary approach to mitigating climate change impacts.[2] Because of their larger historical GHG emissions, developed countries (those included in Annex I of the UNFCCC), are expected to act first under the principle of "common but differentiated responsibilities."[3] Since the signing of the UNFCCC nearly 20 years ago, parties have met periodically to try to bring these principles into action.

Attendees at the first Conference of the Parties (COP) in 1995 launched a round of negotiations resulting in the Berlin Mandate.[4] This mandate ignited the process with a goal to establish more-formal commitments for Annex I countries. Then in December 1997 at COP 3 (third session) in Kyoto, Japan, the UNFCCC moved toward binding commitments for GHG emissions cuts by Annex I countries. These commitments came in the form of the Kyoto Protocol, entered into force in 2005. The Kyoto Protocol established a first compliance period from 2008 to 2012. Notably, the United States did not sign the Kyoto Protocol citing the lack of binding commitments from key developing countries such as China and India.[5]

Parties with commitments under the Kyoto Protocol have accepted targets for limiting or reducing emissions. These targets are expressed as levels of allowed emissions, or "assigned amounts," over the 2008 to 2012 commitment period. The allowed emissions are divided into "assigned amount units" (AAUs).[6]

---

1. United Nations Framework Convention on Climate Change, *opened for signature* May 9, 1992, S. TREATY DOC. No. 102-38, 1771 U.N.T.S. 107 (1992), *available at* http://unfccc.int/essential-background/convention/background/items/2853.php [hereinafter UNFCCC].

2. *See* Kyoto Protocol to the United Nations Framework Convention on Climate Change art. 3.3, Dec. 10, 1997, 37 I.L.M. 22, U.N. Doc. FCCC/CP/1997/7/Add.l (1998) [hereinafter Kyoto Protocol].

3. *Id.* at art. 3.1.

4. United Nations Framework Convention on Climate Change, Mar. 28–Apr. 7, 1995, Berlin, *Report of the Conference of the Parties on its First Session*, U.N. Doc. FCCC/CP/1995/7/Add. 1 (June 6, 1995), *available at* http:/unfccc.int/resource/docs/cop1/07a01.pdf.

5. S. Res. 98, 105th Cong. (1997).

6. Kyoto Protocol, *supra* note 2, art. 3, § 1.

A core component of the Kyoto Protocol is its requirement that countries limit and reduce their GHG emissions or face compliance penalties. By setting such targets, emissions reductions could have an economic value. To help countries meet their emission targets, and to encourage the private sector and developing countries to contribute to reduction efforts, Kyoto incorporates three market-based mechanisms—emissions trading,[7] the clean development mechanism (CDM),[8] and joint implementation (JI).[9] The European Union and its member states were the first to adopt carbon trading systems and participate in CDM and JI projects in order to satisfy Kyoto commitments.[10] This is commonly referred to as the "carbon market."

Emissions trading, as set out in Article 17 of the Kyoto Protocol, allows countries that have emission units to spare—emissions permitted to them but not used—to sell this excess capacity to countries that are over their targets. CDM and JI, on the other hand, allow parties to earn credits toward meeting their Kyoto target by initiating emissions-reduction programs.[11] Under the CDM, parties earn certified emissions-reduction credits (CERs) by initiating emissions-reduction projects in developing countries.[12] Under the JI program, parties do the same in countries subject to a Kyoto emissions reduction or limitation, earning themselves emissions-reduction units (ERUs).[13] In each case, one ERU or one CER is equivalent to one ton of CO_2, and in each case, parties can trade these credits in the carbon market.[14]

The United Nations touts the success of the JI and CDM. "Joint implementation offers Parties a flexible and cost-efficient means of fulfilling a part of their Kyoto commitments, while the host Party benefits from foreign investment and technology transfer."[15] Likewise, the CDM provides flexibility while also stimulating development in countries that need it.[16] As of November 8, 2011, the CDM reports more than 3,500 emissions-reduction projects and more than 770 million issued certified ERUs.[17]

---

7. *Id.* art. 17.

8. *Id.* art. 12.

9. *Id.* art. 6.

10. *See, e.g.,* Directive 2003/87/EC of the European Parliament and of the Council of 13 October 2003 on Establishing a Scheme for Greenhouse Gas Emission Allowance Trading Within the Community and Amending Council Directive 96/61/EC and later amendments, *available at* http://eur-lex.europa.eu/LexUriServ/LexUriServ.do?uri=CONSLEG:2003L0087:20090625:EN:PDF.

11. *Mechanisms Under the Kyoto Protocol: Emissions Trading, the Clean Development Mechanism and Joint Implementation,* UNFCCC, http://unfccc.int/kyoto_protocol/mechanisms/items/1673.php (last visited Aug. 17, 2012).

12. *Clean Development Mechanism (CDM),* UNFCCC, http://unfccc.int/kyoto_protocol/mechanisms/clean_development_mechanism/items/2718.php (last visited Nov. 8, 2011).

13. *Joint Implementation,* UNFCCC, http://unfccc.int/kyoto_protocol/mechanisms/joint_implementation/items/1674.php (last visited Aug. 17, 2012).

14. *Mechanisms Under the Kyoto Protocol, supra* note 11.

15. *Joint Implementation, supra* note 13.

16. *Clean Development Mechanism (CDM), supra* note 12.

17. *CDM in Numbers,* UNFCCC, http://cdm.unfccc.int/Statistics/index.html (last visited Nov. 8, 2011).

Carbon markets, discussed in more detail in Chapter 6, Section F.2, present an interesting issue for insurers. In connection with their participation in a carbon trading market, some insureds may face business interruption problems, contractual liability, errors and omissions claims, and fines and penalties, but such risks are often already addressed in the underwriting process through policy triggers, limitations, and exclusions. For one, the liability-triggering event probably would result in economic damage, rather than property damage. Fines and penalties and contractual liability may be excluded. New political risk–type products may assist participants in market-based mechanisms in managing the associated risks.

Over the past few years, nations have been negotiating approaches for a post-2012 commitment regime under the UNFCCC and the soon-to-expire Kyoto Protocol. These negotiations have led to a series of action plans and agreements at several COP meetings. The goal of these meetings was to bring the United States and key developing countries, such as China, under a binding commitment regime.[18] This goal was consistent with the UNFCCC guiding principle of common but differentiated responsibilities. For example, talks have encouraged developing nations to adopt "nationally appropriate" commitments to reduce emissions.

In large measure, this goal has not been met. Nonetheless, several recent COP meetings have sketched the contours of what many hope to be a comprehensive and vigorous attempt to reduce emissions, mitigate damages, and boost capacity building in developing nations. For example, at COP 13 in Bali, Indonesia, members emphasized that deep cuts in global GHG emissions will be required to achieve the ultimate goals of the UNFCCC.[19] The cornerstones of the Bali Action Plan were:

- a shared vision for long-term cooperative action toward emissions reductions;
- enhanced national and international action on mitigation of climate change that is measurable, reportable and verifiable;
- enhanced action on adaptation;
- enhanced action on technology development and transfer to support action on mitigation and adaptation; and
- enhanced action to provide financial resources and investment that support mitigation and adaptation through technology cooperation.[20]

---

18. Sheila M. Olstead & Robert N. Stavins, *Three Pillars of Post-2012 International Climate Policy*, Harv. Project on Int'l Climate Agreements, http://belfercenter.ksg.harvard.edu/files/stavins_olmstead%20_viewpoint.pdf (last visited Nov. 7, 2011).

19. Conference of the Parties, Fifteenth Session, Bali, Dec. 3–15, 2007, *Bali Action Plan*, U.N. Doc. FCCC/CP/2007/6/Add.1 (Mar. 14, 2008), *available at* http://unfccc.int/resource/docs/2007/cop13/eng/06a01.pdf.

20. *Id.*

COP 14, held in Poznán, Poland, in December 2008, was the important halfway mark to reaching a post-2012 commitment regime.[21] The parties to the UNFCCC reviewed the progress made in 2008 towards a post-2012 commitment regime and mapped out what needed to be accomplished to meet the December 2009 deadline.[22] In addition, the UNFCCC Adaptation Fund was finalized, which establishes adaptation projects for developing countries that are parties to the Kyoto Protocol.[23]

Although critics complained of its lack of action, COP 15 in Copenhagen was a pivotal point in the international effort to reduce carbon emissions. According to the United Nations, the Copenhagen Climate Change Conference in December 2009 "raised climate change policy to the highest political level."[24] The result of this meeting was the Copenhagen Accord.[25] This agreement included a goal of limiting the increase in global temperatures to less than 2 degrees Celsius. Nonetheless, the Copenhagen Accord provided no practical methods to achieving the 2-degrees goal.

In November 2010, the United Nations held COP 16 in Cancun, Mexico, out of which came the Cancun Agreements. These agreements not only reiterated the goal of a maximum rise of 2 degrees Celsius but also articulated concrete steps to reach that goal.[26] For example, participating governments agreed to establish a Green Climate Fund that will support emissions-reducing initiatives in developing countries. Parties also agreed to support climate-friendly technologies. But the United Nations noted that all the promises combined resulted in "only 60% of the emission reductions needed for a 50% chance of keeping temperatures below that goal."[27] More importantly, the fate of the Kyoto Protocol, set to expire in 2012, remained unclear.

The most recent COP meeting, in Durban, South Africa, addressed the Kyoto Protocol, and the parties came to an agreement about a future climate treaty. Out of COP 17 came the Durban Platform for Enhanced Action.[28] Remarkably, the parties

---

21. *See Poznań Climate Change Conference*, UNFCCC, http://unfccc.int/meetings/cop_14/items/4481.php (last visited Nov. 7, 2011).

22. *Id.*

23. *Id.*

24. *Copenhagen Climate Change Conference*, UNFCCC, http://unfccc.int/meetings/copenhagen_dec_2009/meeting/6295.php (last visited Nov. 7, 2011).

25. Conference of the Parties, Fifteenth Session, Copenhagen, Dec. 7–19, 2009, *Copenhagen Accord*, U.N. Doc. FCCC/CP/2009/11/Add.1, Decision 2/CP.15 (Mar. 30, 2010), *available at* http://unfccc.int/resource/docs/2009/cop15/eng/11a01.pdf#page=4.

26. *Cancun Agreements*, UNFCCC, http://unfccc.int/meetings/cancun_nov_2010/items/6005.php (last visited Nov. 7, 2011).

27. *Cancun Climate Change Conference*, UNFCCC, http://unfccc.int/meetings/cancun_nov_2010/meeting/6266.php (last visited Nov. 7, 2011).

28. Draft Decision/CP.17, Advance Unedited Version, http://unfccc.int/files/meetings/durban_nov_2011/decisions/application/pdf/cop17_durbanplatform.pdf (last visited Dec. 20, 2011).

agreed to adopt a universal legal agreement on climate change by 2015.[29] Most importantly, the United States, China, and India signed on to this platform, making it the first time the three biggest emitters have done so.[30] Nonetheless, critics continue to argue the agreement does not go far enough.[31]

Despite the critics, the United Nations process to reduce global carbon emissions is still influential. It continues to raise the profile of climate change issues in the media and therefore among the public. The next COP meeting, COP 18 in Qatar in 2012, will probably begin to develop the contours of a legally binding treaty. The politics could change dramatically, however, between now and 2015. International climate-change regulation, if it ever became more concrete, also could lead to a surge in activity in other climate change contexts. The development of data and guidelines through the United Nations process also could contribute to the development of a "standard of care" in legal disputes related to climate change.

B. National Regulation

1. EPA Action Pursuant to the Clean Air Act

In the United States, EPA has taken steps to use its existing authority under the Clean Air Act to regulate climate change. Although the Bush administration resisted demands to use the Clean Air Act in the climate change context, a significant 2007 Supreme Court decision, *Massachusetts v. EPA*, and the change in administrations paved the way for EPA to use the Clean Air Act to manage GHG emissions. Many commentators, however, have argued that trying to use the 1990 Clean Air Act, the most recent version of the clean air law first passed in 1970, to address climate change is like trying to fit a square peg in a round hole. In fact, in 2011, two bills introduced in the House of Representatives, the No More Excuses Energy Act and the Energy Tax Prevention Act, called for ending EPA's ability to regulate GHG emissions under the Clean Air Act. Neither of these bills has been passed by both the House and the Senate,[32] and until Congress adopts an alternative scheme, EPA is likely to lead the way in developing federal GHG emissions regulation. EPA has continued to regulate

---

29. *Id.* ¶¶ 2–4; *see also Durban Climate Change Conference - Nov. 2011*, UNFCCC, http://unfccc.int/meetings/durban_nov_2011/meeting/6245.php (last visited Dec. 20, 2011).

30. Louise Gray, *Durban Climate Change: The Agreement Explained*, Telegraph (Dec. 20, 2011), http://www.telegraph.co.uk/earth/environment/climatechange/8949099/Durban-climate-change-the-agreement-explained.html.

31. *Id.*

32. The No More Excuses Energy Act did not make its way out of House committees, and the Energy Tax Prevention Act was passed by the House in April of 2011 but not by the Senate. *See* Columbia Law School Center for Climate Change Law, *Climate Legislation Tracker*, H.R. 1023, H.R. 910, http://www.law.columbia.edu/centers/climatechange/resources/legislation (last updated May 3, 2011).

a. *Massachusetts v. EPA*

Massachusetts v. EPA is an important case in the history of GHG litigation and regulation. In *Massachusetts v. EPA*,[33] a group of states, local governments, and private citizens filed a petition asserting that EPA had not properly exercised its responsibility under the Clean Air Act to regulate the emissions of certain GHGs, including CO_2. The questions posed by the plaintiffs in *Massachusetts* were (1) whether EPA has authority to regulate GHG emissions from new motor vehicles pursuant to section 202(a)(1) of the Clean Air Act and, (2) if so, whether EPA's stated reasons for refusing to so regulate are consistent with the Clean Air Act.[34] The plaintiffs asserted that by exercising its regulatory authority, EPA could help reduce GHG emissions from motor vehicles and therefore mitigate global climate change.[35]

Section 202(a)(1) of the Clean Air Act states:

> The [EPA] Administrator shall by regulation prescribe (and from time to time revise) in accordance with the provisions of this section, standards applicable to the emission of any *air pollutant* from any class or classes of new motor vehicles or new motor vehicle engines, which in his judgment cause, or contribute to, air pollution which may reasonably be anticipated to endanger public health or welfare. . . .[36]

EPA argued that it did not possess the power to regulate GHG emissions standards for new motor vehicles pursuant to section 202 of the Clean Air Act. EPA contended that climate change had its own "political history" of putting the issue within the purview of Congress, not EPA.[37] EPA went on to assert that because regulating GHG emissions standards would have wide-ranging political and economic repercussions, EPA was convinced it was devoid of power to regulate in that arena because such power was relegated to Congress. EPA contended that it naturally followed that GHGs could not be "air pollutants under the [Clean Air Act]."[38]

The court rejected EPA's arguments and noted that "[w]hile the Congresses that drafted § 202(a)(1) might not have appreciated the possibility that burning fossil fuels could lead to global warming, they did understand that without regulatory flexibility,

---

33. 549 U.S. 497 (2007).
34. *Id.* at 505.
35. *Id.* at 523.
36. 42 U.S.C. § 7521(a)(1) (2012).
37. Massachusetts v. EPA, 549 U.S. at 512.
38. *Id.* at 513.

changing circumstances and scientific developments would soon render the Clean Air Act obsolete. The broad language of § 202(a)(1) reflects an intentional effort to confer the flexibility necessary to forestall such obsolescence."[39] The court held that GHGs (including CO_2, CH_4, N_2O, and HFCs) fit well within the Clean Air Act's broad definition of "air pollutant," and EPA consequently has the statutory authority to regulate the emission of such gases from new motor vehicles.[40] The court further held that "[u]nder the clear terms of the Clean Air Act, EPA can avoid taking further action only if it determines that greenhouse gases do not contribute to climate change or if it provides some reasonable explanation as to why it cannot or will not exercise its discretion to determine whether they do."[41]

Although the Supreme Court's holding that certain GHGs are "air pollutants" for purposes of section 202(a)(1) of the Clean Air Act is significant, the Supreme Court's ruling in *Massachusetts* does not go as far as holding that GHGs are air pollutants as a general matter. That is, the Supreme Court's decision does not necessarily mean that courts will hold that such GHGs are pollutants for purposes of an insurance policy's pollution exclusion. That said, the Supreme Court's opinion provides insurers with an additional argument that GHGs are pollutants as defined in an insurance policy and fall within a pollution exclusion. See Chapter 8.C.1, further discussing the interpretation of the term "pollutant" in the insurance context and Chapter 8.C.2, for a discussion about the pollution exclusion.

• •

Court rulings and regulatory actions related to climate change could have impacts on the interpretation of the term "pollutant" in the insurance context.

• •

b. EPA Endangerment and Cause or Contribute Findings for GHGs

In response to the Supreme Court's ruling in *Massachusetts v. EPA* and pursuant to section 202(a) of the Clean Air Act, EPA evaluated whether emissions of GHGs from new motor vehicles or new motor-vehicle engines cause, or contribute to, air pollution that may reasonably be anticipated to endanger public health or welfare, or whether the science is too uncertain to make a reasoned decision.[42] On December 7, 2009, EPA Administrator Lisa Jackson signed an Endangerment Finding and a Cause or

---

39. *Id.* at 532.
40. *Id.*
41. *Id.* at 533.
42. *See* Endangerment and Cause or Contribute Findings for Greenhouse Gases Under Section 202(a) of the Clean Air Act, 74 Fed. Reg. 66,496 (Dec. 15, 2009) [hereinafter Endangerment and Cause or Contribute Findings for GHGs] (codified at 40 C.F.R. ch. 1).

Contribute Finding, which set no requirements but was a prerequisite to regulation of GHGs from motor vehicles.[43] First, relying on IPCC, NRC, and USGCRP reports, EPA found that the mix of six key GHGs may be reasonably anticipated to endanger the public health and welfare.[44] In issuing the Endangerment Finding, EPA defined the "air pollution" referred to in section 202(a) of the Clean Air Act as the "mix" of six GHGs: CO_2, CH_4, N_2O, HFCs, PFCs, and SF_6.[45] Interestingly, the defined mix includes GHGs not emitted by motor vehicles (SF_6 and PFCs) and does not include the full universe of GHGs. For scientific and policy reasons, EPA chose not to include water vapor, CFCs, HCFCs, halons, O_3, black carbon, fluorinated ethers, and NF_3 in the definition of air pollution pursuant to section 202(a) of the Clean Air Act.[46] This may have implications for the definition of pollutant in the insurance context. See Chapter 8.C.1.

Concurrently with the Endangerment Finding, the administrator found that the combined emissions of CO_2, CH_4, N_2O, and HFCs from new motor vehicles and new motor-vehicle engines[47] contribute to air pollution that may be reasonably anticipated to endanger public health or welfare.[48] For purposes of the Cause or Contribute Finding, EPA defined the "air pollutant" as the collective class of the six key GHGs rather than individual GHGs.[49] The Cause or Contribute Finding does note, however, that EPA believes it has the discretion to set standards that either control the emissions of the group of GHGs as a whole and/or control emissions of individual GHGs as constituents of the class.[50]

---

43. *Id.*
44. *Id.* at 66,497; *see also* EPA, Technical Support Document for Endangerment and Cause or Contribute Findings for Greenhouse Gases under Section 202(a) of the Clean Air Act 31(Dec. 7, 2009), *available at* http://www.epa.gov/climatechange/Downloads/endangerment/Endangerment_TSD.pdf.
45. Endangerment and Cause or Contribute Findings for GHGs, 74 Fed. Reg. at 66,497.
46. *Id.* at 66,519–21.
47. The Clean Air Act defines "new motor vehicle" and "new motor vehicle engine" as follows: "Except with respect to vehicles or engines imported or offered for importation, the term 'new motor vehicle' means a motor vehicle the equitable or legal title to which has never been transferred to an ultimate purchaser; and the term 'new motor vehicle engine' means an engine in a new motor vehicle or a motor vehicle engine the equitable or legal title to which has never been transferred to the ultimate purchaser; and with respect to imported vehicles or engines, such terms mean a motor vehicle and engine, respectively, manufactured after the effective date of a regulation issued under Section 7521 of this title which is applicable to such vehicle or engine (or which would be applicable to such vehicle or engine had it been manufactured for importation into the United States)." 42 U.S.C. § 7550(3) (2012). Thus, this regulation only relates to new vehicle model years. Older motor vehicles are only affected in the sense that plaintiffs lawyers might use the revelations in these rules to argue that older motor vehicles are defectively designed because they emitted GHGs. *See* Chapter 6.A.6.a, discussing potential product liability theories.
48. Endangerment and Cause or Contribute Findings for Greenhouse Gases, 74 Fed. Reg. 66,496.
49. *Id.*
50. *Id.* at 66,497.

Immediately after Administrator Jackson signed the Endangerment Finding, however, state and industry initiated a series of lawsuits challenging the finding. Most of the suits alleged that EPA should have engaged in a cost-benefit analysis before issuing the Endangerment Finding for GHGs. The suits also challenged (1) the adequacy of the science supporting the Endangerment Finding, (2) EPA's decision not to "quantify" the risk of endangerment to public health or welfare created by climate change, and (3) EPA's choice to define the air pollutant at issue as an aggregate mix of six GHGs.[51] The litigants also disputed EPA's reliance on outside parties, such as IPCC, NRC, and USGCRP, for the scientific data. In 2010, these suits were consolidated into one case, *Coalition for Responsible Regulation v. EPA*, in the D.C. Circuit.

On June 26, 2012, the D.C. Circuit upheld EPA's Endangerment Finding and denied or dismissed the remainder of the petitions for review.[52] The D.C. Circuit held that section 202(a)(1) of the Clean Air Act governing endangerment findings requires scientific judgment, not policy discussions.[53] The statute speaks in terms of endangerment, not policy.[54] Thus, EPA was not required or even permitted to conduct a cost-benefit analysis in the context of the GHG Endangerment Finding.[55]

Significantly, the court also held that the scientific record was adequate to support the Endangerment Finding. The D.C. Circuit concluded that the "body of scientific evidence marshaled by EPA in support of the Endangerment Finding is substantial."[56] The court held that "EPA's scientific record included support for the proposition that greenhouse gases trap heat on Earth that would otherwise dissipate into space; that this 'greenhouse effect' warms the climate; that human activity is contributing to increased atmospheric levels of greenhouse gases; and that the climate system is warming."[57] Based on this evidence, EPA made the "linchpin finding" that the root cause of recently observed climate change is very likely the observed increase in anthropogenic GHG emissions.[58]

Interestingly, the court rejected petitioners' claims that EPA had inappropriately delegated its scientific judgment to the IPCC, USGCRP, or NRC because of EPA's reliance on these "assessments."[59] The court held that relying on peer-reviewed assessments synthesizing thousands of underlying studies was appropriate. EPA

---

51. Coalition for Responsible Regulation v. EPA, No. 09-1322, slip op. at 22 (D.C. Cir. June 26, 2012) (also discussing additional challenges).
52. *Id.*
53. *Id.* at 23.
54. *Id.* at 25.
55. *Id.*
56. *Id.* at 28.
57. *Id.* at 28–29.
58. *Id.* at 29.
59. *Id.* at 26–28.

appropriately sought out the scientific literature and relied on it to form EPA's own judgment.

The court also rejected the petitioners' argument that the Endangerment Finding was indefensible because there is some residual scientific uncertainty.[60] In the court's view, the Clean Air Act does not require certain endangerment.[61] EPA also did not need to quantify the danger in precise numeric terms. The failure to quantify the risk reflected the sometimes uncertain nature of climate science and was not a sign of arbitrary and capricious decision-making.

The petitioners also lacked standing to challenge EPA's decision to make the aggregate of six key GHGs the subject of the Endangerment Finding. Petitioners claimed EPA's decision to include PFCs and SF_6 was arbitrary and capricious. As EPA's decision to regulate those two substances does not injure any motor vehicle–related petitioner or other petitioner, the D.C. Circuit concluded petitioners had no standing to make that challenge.[62]

2. EPA and NHTSA Light-Duty Vehicle GHG Standards and Corporate Average Fuel Economy Standards ("Tailpipe Rule")

As a follow-up to its Endangerment Finding, EPA, in conjunction with the National Highway Traffic Safety Administration Board (NHTSA), established a national program to reduce GHG emissions from light-duty vehicles including passenger cars, light-duty trucks, and medium-duty passenger vehicles. EPA finalized nationwide GHG emissions standards under the Clean Air Act, and NHTSA finalized corporate average fuel economy (CAFE) standards under the Energy Policy and Conservation Act for light-duty vehicles.[63] The new standards establish CO_2 emissions for 2012 through 2016 model years.[64] Impacted vehicles are subject to varying compliance levels based on vehicle size,[65] and the industry is required to meet a fleet-wide average emission level of 250 grams of CO_2 per mile (equivalent to 35.5 miles per gallon) by model year 2016.[66] The rule sets emissions limits for N_2O (0.010 grams per mile) and CH_4 (0.030 grams per mile) for model year 2012 and later,[67] as well as provides

---

60. *Id.* at 31–32.
61. *Id.* at 31.
62. *Id.* at 34.
63. Light-Duty Vehicle Greenhouse Gas Emissions Standards and Corporate Average Fuel Economy Standards, 75 Fed. Reg. 25,324 (May 7, 2010) (codified at 40 C.F.R. pts. 85, 86, 600, 49 C.F.R. §§ 531, 533, 536–38).
64. *Id.*
65. *Id.*
66. *Id.* at 25,369.
67. *Id.* at 25,399.

car manufacturers with incentives to reduce HFC use in air-conditioning systems.[68] Industry groups, companies, and states challenged the rule in *Coalition for Responsible Regulation v. EPA*, No. 10-1092 (D.C. Cir. May 7, 2010), arguing that EPA acted arbitrarily and capriciously in failing to justify and consider the cost impacts of its conclusion that the rule triggers stationary-source regulation under the Prevention of Significant Deterioration Program (PSD) and Title V provisions of the Clean Air Act.[69] The D.C. Circuit upheld that Tailpipe Rule on June 26, 2012.[70]

Meanwhile, EPA is working on proposed standards for the 2017 to 2025 model-year light-duty vehicles.[71] NHTSA and EPA also adopted the first-ever GHG standards for heavy-duty vehicles in 2011.[72]

3. EPA Timing Rule

Pursuant to a longstanding interpretation of the Clean Air Act, EPA confirmed in what has been named the GHG "Timing Rule" that GHGs became "subject to regulation" under the Clean Air Act once EPA enacted the Tailpipe Rule discussed above.[73] This would trigger the Title V and PSD permitting requirements.[74] Through the Timing Rule, EPA delayed the applicability of these permitting requirements until at least January 2, 2011, when the Tailpipe Rule took effect. After litigants challenged this and other EPA GHG rules, the D.C. Circuit upheld EPA's interpretation of the Clean Air Act PSD and Title V permitting triggers and denied the remainder of petitioners' challenge to the Timing Rule.[75] The court held that petitioners lacked standing to challenge the Timing Rule because there was no injury.[76] The permitting requirements were triggered by operation of the Clean Air Act, not the Timing Rule. The Timing Rule only clarified that the permitting requirements would be triggered when the Tailpipe Rule took effect as opposed to when it was enacted. The operation of the Clean Air Act would have triggered the permitting requirements regardless of the Timing Rule. Moreover, the court held that vacatur of the Timing Rule would have no practical effect.[77] See also Chapter 5.B.5 discussing the litigation and the Tailoring Rule.

---

68. *Id.*
69. Coalition for Responsible Regulation v. EPA, No. 09-1322, slip op. at 39 (D.C. Cir. June 26, 2012).
70. *Id.* at 39–45.
71. *See Regulations & Standards, Transportation & Climate*, EPA, http://www.epa.gov/otaq/climate/regulations.htm.
72. Greenhouse Gas Emissions Standards and Fuel Efficiency Standards for Medium- and Heavy-Duty Engines and Vehicles, Final Rule, 76 Fed. Reg. 57,106 (Sept. 15, 2011), updated by 76 Fed. Reg. 65,971 (Oct. 25, 2011).
73. Reconsideration of Interpretation of Regulations That Determine Pollutants Covered by Clean Air Act Permitting Programs, 75 Fed. Reg. 17,004 (Apr. 2, 2010).
74. *Id.*
75. Coalition for Responsible Regulation v. EPA, No. 09-1322, slip op., (D.C. Cir. June 26, 2012).
76. *Id.* at 73–81.
77. *Id.* at 74.

4. EPA's GHG Reporting Rule

As noted in Chapter 4.B.1, on September 22, 2009, EPA issued a final rule requiring targeted facilities to submit certified annual reports detailing emissions of CO_2, CH_4, N_2O, SF_6, HFCs, PFCs, and other fluorinated gases (e.g., NF_3 and HFE). See Chapter 4.B.1, which discusses general requirements and impacted entities. In 2010, EPA amended the rule to require additional types of facilities to monitor and report their GHG emissions. This rule could lead to Clean Air Act liability issues for covered entities as well as their consultants who assist in the preparation of the required reports. Possible violations subject to administrative and civil enforcement include the failure to report, to collect data needed to calculate emissions, to continuously monitor and test as required, to calculate emissions according to the EPA methodologies, and to keep records.[78] These violations could result in civil penalties of up to $37,500 per day.[79] Each day of a violation constitutes a separate violation. EPA can also seek injunctive relief to compel compliance with the rule.

Knowing violations, such as the falsification of records, are subject to criminal fines and possible imprisonment of up to five years.[80] EPA has indicated that because a purpose of the rule is to collect accurate data for program and policy purposes, it intends to be flexible in enforcing the rule, using a variety of approaches to achieve compliance.[81] This may include less punitive measures, such as warning letters or violation notices with opportunities to correct reports.

GHG emissions reports must be self-certified by a "designated representative."[82] The designated representative can be any person, including a third-party contractor, that a company chooses to certify the report and bind the company. It does not have to be the facility manager or a person in senior management. The designated representative must attest under penalty of law that he or she is authorized to submit the report, has personally examined the report, and is familiar with the statements and information submitted, and that the report is accurate and complete to the best of his or her knowledge.[83]

Penalized targeted emitters are likely to turn on their consultants if the emissions reports turn out to be incorrect, which could have implications for professional liability policies if insured consultants were used to gather the relevant data. There is no clear standard for professionals inventorying emissions, but the EPA reporting system will contribute to the development of a "standard of care" for measuring and reporting

---

78. Mandatory Reporting of Greenhouse Gases, 74 Fed. Reg. 56,260, 56,360 (Oct. 30, 2009).
79. Id.
80. See id.; see also Clean Air Act, 42 U.S.C. § 7413(c)(1) (2012).
81. Mandatory Reporting of Greenhouse Gases, 74 Fed. Reg. at 56,361.
82. Id. at 56,355.
83. Id.

GHG emissions. Facilities that are already monitoring emissions for other purposes, e.g., measuring sulfur dioxide under the Acid Rain Program using a continuous emissions monitoring system (CEMS), must upgrade their equipment to monitor GHG emissions.[84] Facilities that do not have a CEMS in place have the option of installing the equipment or calculating GHG emissions according to the methodologies EPA has specified for each industrial sector covered by the rule.[85] There is no universal methodology because the parameters that are used in the calculations differ for each industrial sector, e.g., cement production versus manure management. Moreover, the technology is still developing. To try to ease the transition, the EPA rule allowed covered sources to use "best available monitoring methods" for the first quarter of 2010.[86] For professional liability, these developments potentially mean both market opportunity and uncertainty.

Reporting for impacted entities could be complicated by the fact that the GHG Reporting Rule does not preempt or replace state reporting programs.[87] Thus, GHG emitters also will be subject to various state reporting requirements, some of which contain more rigorous standards than the federal rule.[88] For example, Washington State recently adopted a reporting rule that applies to several source types not covered by the federal rule; reduces the emissions threshold for reporting to 10,000 metric tons starting in 2012 (as opposed to 25,000 metric tons under the federal rule); requires reporting of indirect emissions (the federal rule does not); and requires third-party verification of submitted reports (the federal rule allows entities to self-certify).[89] Many state reporting programs are coupled with regional climate-change agreements, such as the Western Climate Initiative, of which Washington is a member. See Chapter 5.G. Thus, although EPA focuses on larger GHG emitters, states could choose to target smaller emitters.

5. EPA Prevention of Significant Deterioration and Title V Greenhouse Gas Tailoring Rule

As noted above, as a consequence of EPA's implementation of light-duty vehicle emissions standards, EPA concluded that GHGs[90] would become pollutants "subject to regulation" under the Clean Air Act, a circumstance that would trigger regulatory

---

84. *Id.* at 56,268.
85. *Id.*
86. *Id.*
87. *See* EPA, GHG Reporting Program Frequently Asked Questions (FAQs), 1.7.37 Q53, *available at* http://www.ccdsupport.com/confluence/download/attachments/3113574/FAQs.pdf?version=9&modificationDate=1322584702000.
88. In fact, EPA specifically notes that "[m]any state programs are broader in scope, in a more advanced state of development, and have different policy objectives than [EPA's] rulemaking." *Id.*
89. Reporting Emissions of Greenhouse Gases, Wash. Admin. Code § 173-441-010 (2011).
90. Although the light-duty vehicle GHG standards apply only to CO_2, N_2O, CH_4, and HFC emissions, EPA also incorporated SF_6 and PFCs into its Tailoring Rule. This suggests that the EPA may develop additional emissions standards under the Clean Air Act applicable to SF_6 and PFCs.

obligations under additional Clean Air Act programs such as the New Source Review (NSR) Prevention of Significant Deterioration (PSD)[91] and Title V[92] permitting programs.[93] Pursuant to those programs, stationary sources emitting more that 100 tons per year (for the Title V program and for specified source categories under the PSD program) or 250 tons per year (for other source categories under the PSD program) of a regulated substance are "major sources" required to obtain permits for new construction or modification of a facility (under the NSR program) or continuously maintain permits while operating a facility (under the Title V program). Typically, permits obtained pursuant to these programs set emissions standards and establish control technology requirements. To obtain a PSD permit, a covered source must install best achievable control technology (BACT) for each pollutant subject to regulation under the Clean Air Act, not just the NAAQS pollutant.

Recognizing that the emissions triggers currently used by the NSR PSD and Title V programs (100 or 250 tons per year) would be unmanageable in the context of GHGs,[94] EPA adopted the Tailoring Rule that sets alternate triggers for the impacted GHGs.[95] EPA decided that permitting requirements for GHGs should only apply to the largest sources and the sources with the most experience with permitting.

Pursuant to the Tailoring Rule, the PSD and Title V programs are only applicable to facilities in two categories: (1) those that emit 75,000 tons or more of GHGs on a carbon dioxide equivalent (CO_2e) basis per year and are already required to report their emissions of non-GHGs and (2) those facilities that are not otherwise required to obtain permits for any other pollutant but emit 100,000 tons or more per year of CO_2e.[96] EPA estimates that even with the increased emissions triggers, approximately

---

91. See 40 C.F.R. § 52.21(b)(50)(iv) (defining a "regulated NSR pollutant" as "[a]ny pollutant that otherwise is subject to regulation under the [Clean Air] Act").

92. See id. § 70.2.

93. Reconsideration of Interpretation of Regulations That Determine Pollutants Covered by Clean Air Act Permitting Programs, 75 Fed. Reg. 17,004 (Apr. 2, 2010).

94. According to the EPA, without this Tailoring Rule the number of PSD permits issued each year would have increased from approximately 280 to more than 41,000 and the number of Title V permits would increase from approximately 14,000 to more than 6 million. EPA, EPA-HQ-OAR-2009-0517, PREVENTION OF SIGNIFICANT DETERIORATION AND TITLE V GREENHOUSE GAS TAILORING RULE, at 50, 56–57 (2009), available at http://www.epa.gov/nsr/documents/GHGTailoringProposal.pdf.

95. Prevention of Significant Deterioration and Title V Greenhouse Gas Tailoring Rule, 75 Fed. Reg. 31,514 (June 3, 2010).

96. For modifications of existing sources, an increase in emissions of 75,000 tons per year would trigger PSD permitting obligations. Prevention of Significant Deterioration and Title V Greenhouse Gas Tailoring Rule, 75 Fed. Reg. 31,514 (June 3, 2010) (codified at 40 C.F.R. 50, 51, 70, 72); EPA, FINAL RULE: PREVENTION OF SIGNIFICANT DETERIORATION AND TITLE V GREENHOUSE GAS TAILORING RULE FACT SHEET, at 2, available at http://www.epa.gov/nsr/documents/20100413fs.pdf. However, when determining whether an increase reaches the 75,000 ton threshold, emissions from fugitive sources (those that do not pass through a chimney, or a stack) are not considered. EPA, PSD and NSR: Reconsideration of Inclusion of Fugitive Emissions, 76 Fed. Reg. 17,548 (Mar. 30, 2011) (codified at 40 C.F.R. pts. 51, 52).

67 percent of national GHG emissions, including those from power plants, refineries, and cement production facilities, are covered by permitting requirements.[97]

In April 2010, EPA issued a rule stating that regulation of GHGs under the Clean Air Act's PSD and Title V programs would apply after January 2, 2011.[98] As a result, for the facilities in the first category, the Tailoring Rule did not take effect until January 2, 2011, and for those in the second category, the rule was only applicable after July 1, 2011.[99] Additionally, for biomass and other biogenic sources, the permitting requirements will not take effect until at least 2014.[100]

In December 2010, EPA issued a series of rules that put the necessary regulatory framework in place to ensure that (1) industrial facilities get Clean Air Act permits covering their GHGs when needed and (2) facilities emitting GHGs at levels below those established in the Tailoring Rule do not need to obtain federal Clean Air Act permits.[101] EPA issued a rule finding that 13 states (Arizona, Arkansas, California, Connecticut, Florida, Idaho, Kansas, Kentucky, Nebraska, Nevada, Oregon, Texas, and Wyoming) did not meet Clean Air Act requirements because their PSD programs did not cover GHGs.[102] EPA also issued a "SIP Call" requiring these states to revise their state implementation plans (SIPs) to ensure that their PSD programs cover GHGs.[103] Seven of the states agreed to respond before December 22, 2010—Arizona, Arkansas, Florida, Idaho, Kansas, Oregon, and Wyoming.[104] Another five jurisdictions agreed to respond after January 2, 2011, but before making any permit decisions in 2011—Kentucky; Clark County, Nevada; Connecticut; parts of California; and Nebraska.[105] Texas

---

97. With the obvious exception of mobile sources, it appears that this rule encompasses many of the same types of sources covered by the GHG reporting rule. Under both rules, large stationary sources emitting over 75,000 tons of CO_2e per year are impacted. The GHG reporting rule, however, may also encompass certain sources emitting less than 75,000 tons.

98. Reconsideration of Interpretation of Regulations That Determine Pollutants Covered by Clean Air Act Permitting Programs, 75 Fed. Reg. 17,004 (Apr. 2, 2010) (codified at 40 C.F.R. pts. 50, 51, 70, 71).

99. EPA, Final Rule: Prevention of Significant Deterioration and Title V Greenhouse Gas Tailoring Rule Fact Sheet, at 2, *available at* http://www.epa.gov/nsr/documents/20100413fs.pdf.

100. On July 20, 2011, EPA issued a final rule officially deferring for a period of three years the application of the Prevention of Significant Deterioration (PSD) and Title V permitting requirements to biogenic CO_2 from bioenergy and other biogenic stationary sources. 76 Fed. Reg. 43,490 (July 20, 2011).

101. EPA, Clean Air Act Permitting for Greenhouse Gas Emissions—Final Rules: Fact Sheet, *available at* http://www.epa.gov/nsr/ghgdocs/20101223factsheet.pdf.

102. Action to Ensure Authority to Issue Permits Under the Prevention of Significant Deterioration Program to Sources of Greenhouse Gas Emissions: Finding of Substantial Inadequacy and SIP Call, 75 Fed. Reg. 77,698, 77,700 (Dec. 13, 2010) (codified at 40 C.F.R. pt. 52).

103. *Id.* at 77,698.

104. EPA, Final Action to Ensure Authority to Issue Permits Under the Prevention of Significant Deterioration Program to Sources of Greenhouse Gas Emissions: Finding of Substantial Inadequacy and SIP Call: Fact Sheet, at 2, *available at* http://www.epa.gov/NSR/documents/20101201factsheet.pdf.

105. *Id.*

did not elect a date to submit a revised SIP.[106] Later in December 2010, EPA issued a rule giving itself authority to permit GHG emissions in the PSD programs in seven states until the identified state and local agencies revised their permitting regulations to cover GHGs as defined by the Tailoring Rule.[107] These are states that missed the December 22, 2010, deadline but agreed to comply with the Tailoring Rule.[108]

EPA also took steps to provide for permitting while Texas was not in compliance.[109] EPA found that the Texas SIP was not in compliance with the new source requirements under the PSD program and the Tailoring Rule.[110] As a result, the EPA administrator signed a final rule that gave EPA control over implementation of permitting under a Federal Implementation Plan (FIP) until Texas complied with the requirements of the Tailoring Rule.[111]

EPA has issued guidance to assist with implementation of GHG-related permitting requirements. In setting PSD standards for a specific GHG emitter, EPA will consider both the emissions limitations achievable at that facility using what EPA has determined to be the best available control technologies and the costs of implementing those technologies.[112] EPA issued BACT guidance for GHGs in March 2011.[113] EPA also has issued a series of BACT GHG white papers for certain industries. EPA, however, likely will issue further rules to implement its regulatory goals.

The Tailoring Rule has not been without challenge. Industry groups and states filed dozens of lawsuits challenging the Tailoring Rule and questioning EPA's authority to adjust the statutorily defined emissions standards. On November 8, 2010, those

---

106. *Id.*
107. Action to Ensure Authority to Issue Permits Under the Prevention of Significant Deterioration Program to Sources of Greenhouse Gas Emissions: Finding of Failure to Submit State Implementation Plan Revisions Required for Greenhouse Gases, 75 Fed. Reg. 81,874 (Dec. 29, 2010) (codified at 40 C.F.R. pt. 52).
108. *See* Letter from Ariz. Dep't Envtl. Quality to EPA (Aug. 27, 2010), *available at* http://www.epa.gov/nsr/2010letters/az.pdf; Letter from Fla. Dep't of Envtl. Prot. to EPA (July 2, 2010), *available at* http://www.epa.gov/nsr/2010letters/fl.pdf; Letter from Ark. Dep't of Envtl. Quality to EPA (July 26, 2010), *available at* http://www.epa.gov/nsr/2010letters/ar.pdf; Letter from Idaho Dep't of Envtl. Quality to EPA (July 28, 2010), *available at* http://www.epa.gov/nsr/2010letters/id.pdf; Letter from Kan. Dep't of Health and Env't to EPA (Aug 2, 2010), *available at* http://www.epa.gov/nsr/2010letters/ks.pdf; Letter from Or. Dep't of Envtl. Quality to EPA (July 29, 2010), *available at* http://www.epa.gov/nsr/2010letters/or.pdf; Letter from Wyo. Dep't of Envtl. Quality to EPA (July 30, 2010), *available at* http://www.epa.gov/nsr/2010letters/wy.pdf.
109. Determinations Concerning Need for Error Correction, Partial Approval and Partial Disapproval, and Federal Implementation Plan Regarding Texas's Prevention of Significant Deterioration Program, 76 Fed. Reg. 25,178 (May 3, 2010) (codified at 40 C.F.R. pt. 52).
110. EPA, Clean Air Act Permitting for Greenhouse Gas Emissions—Final Rules: Fact Sheet, at 2, *available at* http://www.epa.gov/NSR/ghgdocs/20101223 factsheet.pdf.
111. *Id.*
112. *Id.*
113. EPA, Office of Air & Radiation, Guidance for Determining Best Available Control Technology for Reducing Carbon Dioxide Emissions From Bioenergy Production, at 5 (2011), *available at* http://www.epa.gov/nsr/ghgdocs/bioenergyguidance.pdf.

cases were consolidated into *Coalition for Responsible Regulation v. EPA*.[114] On June 26, 2012, the D.C. Circuit held that petitioners lacked standing to challenge the Tailoring Rule because they could not articulate an injury.[115] The operation of the Clean Air Act, after enactment of the Tailpipe Rule, triggered the permit requirements, not the Tailoring Rule. The court held that, if anything, the Tailoring Rule tempered the operation of the statute.

Certain states filed additional lawsuits related to a series of EPA rules ordering states to revise their Clean Air Act SIPs to accommodate the new GHG regulation. For example, after EPA found Texas's permitting plan inadequate to address GHGs, Texas filed a petition for review in the D.C. Circuit challenging EPA's authority.[116] Other such petitions were consolidated as *Utility Air Regulatory Group v. EPA*, No. 11-1037 (D.C. Cir.). These challenges may be decided in 2012.[117]

Lawsuits also are likely to arise out of specific grants or denials of PSD and Title V permits as the implementation of the Tailoring Rule and other Clean Air Act programs goes forward.[118] For example, there could be debate over what qualifies as BACT for a particular industry. Permit challenges have been frequent outside the GHG context.

6. Standards of Performance for Carbon Dioxide Emissions for New Stationary Sources: Electric Utility Generating Units

On March 27, 2012, EPA proposed Standards of Performance for Greenhouse Gas Emissions for New Stationary Sources: Electric Utility Generating Units.[119] The rule is based in part on EPA's 2009 Endangerment Finding with respect to GHGs.

The proposed rule applies only to carbon dioxide emissions from *new* fossil-fuel-fired electric utility generating units (EGUs). Under the rule, fossil-fuel-fired EGUs include fossil-fuel-fired boilers, integrated gasification combined cycle (IGCC) units,

---

114. Coalition for Responsible Regulation v. EPA, No. 10-1073 (D.C. Cir.) (consolidating cases challenging the Timing Rule and Tailoring Rule). Note that although this case has the same first named party, it is a separate case from the *Coalition for Responsible Regulation v. EPA*, No. 09-1322, discussed in Chapter 5.B.1.b.

115. Coalition for Responsible Regulation v. EPA, No. 09-1322, slip op., at 73–82 (D.C. Cir. June 26, 2012).

116. Texas v. EPA, No. 10-1425 (D.C. Cir. filed Feb. 11, 2010).

117. In the meantime, new power plant projects may be delayed or derailed altogether. *See* Matthew Tresaugue, *First Casualty of Greenhouse Gas Rules May Be Texas Plant*, Hous. Chron. (July 6, 2012), http://www.chron.com/news/politics//news/houston-texas/article/First-casualty-of-greenhouse-gas-rules-may-be-3689643.php.

118. EPA must publish notice of any PSD permit decision in the *Federal Register* and provide a time frame (usually 30 days) for public comment. After a permitting decision becomes final, it is subject to judicial review. In the past, EPA grants of PSD approvals often have led to litigation. *See, e.g., In re* Deseret Power Elec. Coop., PSD Appeal No. 07-03, 2008 WL 5572891 (EAB Nov. 13, 2008); *Longleaf Energy Assocs., LLC v. Friends of the Chattahoochee, Inc.*, 681 S.E.2d 203 (Ga. Ct. App. 2009).

119. *See* Standards of Performance for Greenhouse Gas Emissions for New Stationary Sources: Electric Utility Generating Units, 77 Fed. Reg. 72 (Apr. 13, 2012); EPA Fact Sheet: Proposed Carbon Pollution Standard for New Power Plants, *available at* http://epa.gov/carbonpollutionstandard/pdfs/20120327factsheet.pdf.

and stationary combined cycle turbine units that generate electricity for sale and are larger than 25 megawatts (MW). The proposal does not apply to plants that burn biomass only or to modifications of existing plants. The proposed rule also does not apply to new power plant units that have permits and start construction within 12 months of this proposal.

EPA is proposing that new fossil-fuel-fired power plants meet an output-based standard of 1,000 pounds of CO_2 per megawatt-hour (lb CO_2/MWh gross). New plants may use carbon capture and sequestration (CCS) to meet the standard. EPA believes the majority of new natural gas combined cycle (NGCC) power plant units will meet the proposed standards without incorporating additional adjustments into construction plans.

This rule is expected to make construction of new coal-fired power plants less likely. The percentage of U.S. electricity obtained from coal-fired power plants already had decreased in recent years due to relatively cheaper, accessible, and plentiful natural gas.[120]

7. Potential Impact of Emerging Clean Air Act Regulation on Climate Change-Related Litigation

As EPA regulates GHGs under the Clean Air Act, litigants are arguing that federal nuisance suits are barred by principles of displacement and state law tort claims are preempted. In *American Electric Power Co. v. Connecticut*, for example, the Supreme Court held that plaintiffs' federal common-law nuisance claims were displaced by federal regulation of GHGs.[121] Courts have not yet had the opportunity to rule on any preemption challenges. If Congress ever enacts comprehensive GHG regulation, this also could result in new arguments about preemption of state law claims.

C. Congressional Action

In 2009 it appeared that climate change-related legislation was within the realm of possibility. On June 26, 2009, the U.S. House of Representatives passed the American Clean Energy and Security Act of 2009 (Waxman-Markey Bill) by a narrow margin of 219 to 212. On September 30, 2009, a similar bill, the Clean Energy Jobs & American Power Act, was introduced in the Senate by Senators Kerry and Boxer (Kerry-Boxer Bill). Neither bill was ever enacted into law but they are examples of the type of legislation we may see again.

---

120. Of course, natural gas prices, accessibility, and stores can and do fluctuate. *See, e.g.,* James M. Inhofe & Frank Fannon, *Energy and Environment: The Future of Natural Gas in America*, 26 ENERGY L.J. 349, 350–52 (2005).

121. Am. Elec. Power Co. v. Connecticut, 131 S. Ct. 2527 (2011).

Congressional activity on climate change is of importance to the insurance industry because climate change-related claims likely would spike after enactment of climate change legislation. Plaintiffs seeking to recover from GHG emitters will see such legislation as imprimatur to their claims and regulated entities might become particular targets in tort and other lawsuits. Legislation, depending on how it is drafted, however, may eventually displace or preempt some claims. The economy and politics affect whether legislation will ever be a reality. The cyclical trends should be monitored.

D. Regulation and Potential Litigation Related to Carbon Storage

Increasing regulation of GHG emissions, coupled with escalating political pressure to decrease the country's carbon footprint, has provoked interest in the development and implementation of new technologies aimed at mitigating climate change. One technology in particular, carbon capture and sequestration, is the focus of both national and international mitigation efforts. That said, the current cost of the technology and permitting factors have stalled all but the smallest of pilots so far.

CCS involves capturing CO_2 as it is emitted from large point sources like fossil fuel power plants, transporting the CO_2 via pipeline to an underground storage location, compressing and injecting the captured gas through injection wells into underground geologic formations for long-term storage. Some insurers have developed a specific set of insurance products targeted at this application of technologies.

The EPA finalized requirements for CCS under the authority of the Safe Drinking Water Act's Underground Injection Control (UIC) Program.[122] The rule creates a new class of injection wells (Class VI) for CO_2 injections and establishes technical criteria for geologic site characterization and well location, well construction and operation, mechanical integrity testing and monitoring, well plugging, post-injection site care, and site closure. EPA will also be tracking the amount of GHGs that are injected. EPA finalized an amendment to the GHG Reporting Rule in December 2010 to ensure the amount of GHGs being injected is tracked.[123]

Various states, including Washington, California, Kansas, Massachusetts, Montana, New Mexico, North Dakota, Oklahoma, Pennsylvania, Texas, Utah, West Virginia, and Wyoming, also have enacted legislation or proposed regulation of CCS,

---

122. Federal Requirements Under the Underground Injection Control (UIC) Program for Carbon Dioxide (CO_2) Geologic Sequestration (GS) Wells, 75 Fed. Reg. 77,230 (Dec. 10, 2010) (codified at 40 C.F.R. pts. 124, 144–47).

123. Mandatory Reporting of Greenhouse Gases: Injection and Geologic Sequestration of Carbon Dioxide; Final Rule, 75 Fed. Reg. 75,060 (Dec. 1, 2010).

meaning that CCS projects may be subject to varying regulations depending on location.[124] As CCS projects proliferate, it is likely that additional states will take regulatory action.

E. Renewable Energy: An Overview

The predominant source of energy in the world is fossil fuels. As concerns about the consequences of increasing GHG emissions from fossil fuel emissions increase, however, demand has increased for alternatives. Alternative energy sources constitute a small but growing share of the overall U.S. and global energy mix. Alternative or renewable energy may include solar, wind, hydropower, geothermal, biofuel, and wave, tide, and current sources. According to the U.S. Energy Information Administration, renewable energy generation constituted 10 percent of total U.S. electricity generation in 2010, and this share is likely to grow to 16 percent by 2035.[125] The non-hydropower renewable share of total U.S. electricity generation is predicted to increase from 4 percent to 9 percent over the same time period.[126] Moreover, non-hydropower renewable generation will account for one-third of the overall growth in electricity generation from 2010 to 2035 due largely to increased wind and biomass.[127] Solar energy is also projected to increase nearly sevenfold by 2035, primarily through a growth in photovoltaics.[128]

Globally, the International Energy Agency (IEA) estimates that renewables and natural gas will constitute nearly two-thirds of incremental energy demand (i.e., any additional energy use above average or normal energy use) between 2010 and 2035.[129] If government policies successfully facilitate growth in the renewable-energy sector, the IEA projects that renewable power could increase from 3,900 terawatt hours (TWh) in 2009 to 11,100 TWh in 2035.[130] Both wind and hydropower would account for approximately one-third each of this growth in renewable generation.[131]

---

124. BUREAU OF NAT'L AFFAIRS, WORLD CLIMATE CHANGE REPORT: THE TROUBLE WITH ANGELS: CARBON CAPTURE AND STORAGE HURDLES AND SOLUTIONS, 3–4 (May 2008).
125. ENERGY INFO. ADMIN., ANNUAL ENERGY OUTLOOK 2012 EARLY RELEASE OVERVIEW 1 (2012), available at http://www.eia.gov/forecasts/aeo/er/pdf/0383er(2012).pdf.
126. *EIA Projects U.S. Non-Hydro Renewable Power Generation Increases, Led by Wind and Biomass*, ENERGY INFO. ADMIN. (Feb. 28, 2012), http://205.254.135.7/todayinenergy/detail.cfm?id=5170.
127. *Id.*
128. *Id.*
129. Int'l Energy Agency, *World Energy Outlook 2011—Presentation to the Press in London* 4 (Nov. 2011), available at http://www.worldenergyoutlook.org/media/weowebsite/2011/WEO2011_Press_Launch_London.pdf.
130. Sonal Patel, *World Energy Outlook Forecasts Great Renewable Growth*, POWER (Jan. 1, 2012), http://www.powermag.com/gas/4244.html.
131. *Id.*

IEA also estimates that biomass and solar photovoltaics could account for around 17 percent and 10 percent, respectively, of all renewable generation.[132]

Energy policies have an extraordinary impact on the portion of energy derived from renewable sources. An increase in renewable generation has been or probably will be driven by a number of factors, including tax incentives for renewable production, proposed and existing greenhouse gas regulations, and renewable portfolio standards (RPS). Renewable producers in the United States can currently qualify for a number of tax breaks, including production and investment tax credits. Section 1603 of the American Recovery and Reinvestment Act (ARRA) also established the option of cash grants in lieu of tax credits program, although the authorization for this program expired at the end of 2011.[133] In addition, the Energy Policy Act of 2005 provided the U.S. Department of Energy (DOE) with authority to offer loan guarantees for innovative clean energy technologies.[134] ARRA modified DOE's loan guarantee authority and established a temporary loan guarantee program for the development and deployment of clean energy technologies and projects.[135]

Congress has not passed legislation that would impose a cap on GHG emissions or establish a federal renewable or clean energy standard. There have, however, been some significant actions at the state and regional levels that might have an effect on the demand for renewables. The most ambitious effort is in California, where the state's economy-wide cap-and-trade program began on January 1, 2012, with compliance obligations starting in 2013.[136] In 2008, 10 Northeastern and Mid-Atlantic states formed the Regional Greenhouse Gas Initiative (RGGI), which established a cap-and-trade system for GHG emissions from the utility sector.[137]

In addition, as of January 2012, 30 states and the District of Columbia have adopted binding renewable energy portfolio standards or other mandated renewable capacity policies.[138] Again, California arguably has established the most aggressive RPS with a requirement that 33 percent of the state's retail sales of electricity be derived from renewable sources by 2020.[139] North Dakota, South Dakota, Utah,

---

132. *Id.*
133. American Recovery and Reinvestment Act of 2009 § 1603, Pub. L. 111-5, 123 Stat. 115 (2009).
134. Phillip Brown, Cong. Research Serv. R42152, Loan Guarantees for Clean Energy Technologies: Goals, Concerns, and Policy Options, 1 (2012).
135. *Id.*
136. *Cap-and-Trade Program*, Cal. Air Res. Bd., (last reviewed Aug. 17, 2012), http://www.arb.ca.gov/cc/capandtrade/capandtrade.htm.
137. Regional Greenhouse Gas Initiative, http://www.rggi.org/. (In 2011, Governor Chris Christie announced that New Jersey was withdrawing from RGGI.)
138. *Most States Have Renewable Portfolio Standards*, Energy Info. Admin. (Feb. 3, 2012), http://205.254.135.7/todayinenergy/detail.cfm?id=4850.
139. *Id.*

Oklahoma, Indiana, Virginia, and Vermont have adopted nonbinding targets for renewable energy in lieu of an RPS.[140]

EPA's regulations of GHG emissions under the Clean Air Act could also impact future demand for renewables. Large, new, or modified stationary sources are required to install BACT to control GHG emissions. In the spring of 2012, EPA proposed new source-performance standards for CO_2 emissions from fossil-fuel-fired EGUs.[141] In addition to GHG emissions, in December 2011, EPA signed a mercury and air toxics standards (MATS) for power plants that will reduce emissions from new and existing coal- and oil-fired EGUs.[142] These regulations may make renewables more cost-competitive in relation to coal and other traditional sources.

Challenges for Renewable Projects. Despite the promise of emerging sources and technologies, there are a number of challenges associated with the development and use of renewable energy that could affect its growth, including:

- *Reliability:* Consumers expect power to be readily available. Renewable sources such as wind and solar, however, are dependent on intermittent weather conditions and may require use of backup power generation resources, such as natural gas.[143]
- *High Capital Costs:* Although renewable power generally is characterized by low operational costs, initial capital costs can be quite high. The construction costs associated with renewable power projects (particularly wind, geothermal, and large solar) can contribute to a perception that these technologies are not as competitive as other kinds of power sources.[144]
- *Siting and Permitting Difficulties:* Public or political opposition to new power lines and renewable energy equipment (e.g., large turbines needed to generate

---

140. *Id.*

141. *See* Standards of Performance for Greenhouse Gas Emissions for New Stationary Sources: Electric Utility Generating Units, 77 Fed. Reg. 22,392 (Apr. 13, 2012); EPA Fact Sheet: Proposed Carbon Pollution Standard for New Power Plants, *available at* http://epa.gov/carbonpollutionstandard/pdfs/20120327factsheet.pdf.

142. National Emission Standards for Hazardous Air Pollutants From Coal and Oil-Fired Electric Utility Steam Generating Units and Standards of Performance for Fossil Fuel-Fired Electric Utility, Industrial-Commercial-Institutional, and Small Industrial-Commercial-Institutional Steam Generating Units, 77 Fed. Reg. 9304 (Feb. 16, 2012).

143. Timothy F. Sutherland, *Building a Clean and Reliable Energy Future: Natural Gas and Renewable Energy*, Nat'l Geographic Online (Mar. 17, 2011), http://www.greatenergychallengeblog.com/2011/03/17/building-a-clean-and-reliable-energy-future-natural-gas-and-renewable-energy/.

144. *Id.; see also* Edison Electric Institute, *Renewable Energy: Growth and Challenges in the Electric Power Industry* 19 (2008).

wind power) can significantly slow development, as can the need for regulatory and environmental assessment.[145]

+ *Financing Difficulties/Lack of Proper Incentives*: Renewable energy project developers also commonly face difficulties obtaining financing to take projects from the research and development stage to widespread commercialization. Financing difficulties are exacerbated by the expiration of the Section 1603 grant program and uncertainty over the future of production tax credits and federal loan guarantees for renewable energy projects. Moreover, the absence of comprehensive climate or renewable energy standard legislation leads to an absence of long-term price signals with regard to the price of carbon and future mandates for renewable production. As with many new technologies, there is hesitancy among investors to provide capital to fund projects when there is a lack of understanding of the technology being employed and the associated risks.

Insurance products may help bridge the financing gap often associated with renewable projects and to address liabilities that may arise in connection with the development, installation, and use of alternative energy. Traditional property and casualty insurance, along with specific new products targeted to the needs of renewable energy projects, like warranty policies, may assist owners, operators, and contractors with risk management. See Chapter 7.C, discussing coverage potentially available for renewable energy projects.

F. State and Local Regulation

> States, and particularly California, have taken an active lead in passing legislation to limit emissions. Such legislation may require insurers to underwrite risks differently depending on the jurisdiction.

Given that the federal legislative branch did not adopt comprehensive GHG legislation, states and localities have acted to set emissions-reduction targets, mandate investment in renewable and efficient-energy sources, develop climate action plans, and design GHG cap-and-trade programs. For example, across the United States, 1,054 mayors have signed their cities on to the U.S. Conference of Mayors Climate

---

145. One notable example of the impact of these kinds of difficulties is the Cape Wind project, which proposes to site 130 large turbines off the coast of Massachusetts in Nantucket Sound. Shalini P. Vajjhala, *Siting Difficulty and Renewable Energy Development: A Case of Gridlock?*, RESOURCES MAG., Winter 2007, at 5.

Protection Agreement.[146] By signing the agreement, a city commits to its own GHG reduction targets and engages in state and federal advocacy campaigns for similar goals. As of June 2012, at least 31 states had completed climate action plans to reduce GHG emissions levels and two other states had plans under way.[147] States that have not taken action are found primarily in coal generation and production regions of the Midwest and Southeast. In addition, as of January 2012, 30 states and the District of Columbia have adopted binding renewable energy portfolio standards or other mandated renewable capacity policies.[148] Seven additional states have voluntary renewable energy goals.[149] See also Chapter 5.E on renewable energy.

California, in particular, has been quite active. As noted, California historically has used its environmental impact review process to regulate climate change. See Chapter 6.F.1.b for a full discussion of the California Environmental Quality Act and related litigation. In addition, California has passed several pieces of legislation in the past decade aimed at reducing GHG emissions. In 2002, California passed Assembly Bill (A.B.) 1493, which required the California Air Resources Board (CARB) to develop and implement GHG limits for vehicles.[150] In 2004, CARB approved regulations limiting the amount of GHGs that can be released from new passenger cars, SUVs, and pickup trucks sold in California.[151] In response to A.B. 1493, the automotive industry filed several lawsuits claiming that the new standards were preempted by federal rules. On December 11, 2007, the U.S. District Court for the Eastern District of California ruled that the state and federal laws could coexist.[152] A subsequent EPA ruling, however, rendered that court's decision moot.

In order to implement the new, more stringent GHG limits for vehicles, California had to apply to EPA for a waiver of the Clean Air Act's preemption of state emissions standards.[153] EPA informed CARB of its intent to deny the board's origi-

---

146. Mayors Climate Protection Center, *Map of Participating Mayors*, http://www.usmayors.org/climateprotection/map.asp (last updated June 11, 2012).

147. Center for Climate Strategies, *State and Local Climate Blackboard*, http://www.climatestrategies.us/policy_tracker/state/index (last visited June 13, 2012).

148. *Most States Have Renewable Portfolio Standards*, ENERGY INFO. ADMIN. (Feb. 3, 2012), http://205.254.135.7/todayinenergy/detail.cfm?id=4850.

149. *Id.*

150. Assemb. B. No. 1493 (Cal. 2002).

151. CAL. CODE OF REGS., tit. 13, § 1961 (2004).

152. *See* Cent. Valley Chrysler-Jeep, Inc. v. Goldstone, No. 1:04-cv-06663-AWI-GSA (E.D. Cal. Dec. 11, 2007), *available at* http://www.climatelaw.org/cases/country/us/case-documents/us/us/Ishii%20Order.pdf.

153. Pursuant to the Clean Air Act, although EPA sets emissions standards for new motor vehicles, California is permitted to adopt its own standards for new motor vehicles if the administrator of EPA waives the general statutory prohibition on state adoption or enforcement of emissions standards. According to section 209 of the Clean Air Act, EPA must grant a waiver unless it finds that California (1) was arbitrary and capricious in its finding that its standards are in the aggregate at least as protective of public health and welfare as applicable to

nal request for a waiver on December 19, 2007—two days after the Eastern District Court ruled that California's standards were not preempted.[154] On March 6, 2008, EPA finalized its determination. On January 26, 2009, however, President Barack Obama signed a Presidential Memorandum directing EPA to assess whether denial of the waiver was appropriate in light of the Clean Air Act,[155] and on June 30, 2009, EPA reversed course and granted a waiver, thus allowing California to implement its GHG emissions standards for motor vehicles beginning with the 2009 model year.[156] Other states may adopt new motor-vehicle emissions standards identical to California's if certain statutory criteria are met. At least 14 states have already adopted California's standards, and additional states have expressed interest in adopting the standards.[157] On September 9, 2009, the U.S. Chamber of Commerce and the National Automobile Dealers Association filed a petition for review of EPA's decision to grant the California waiver.[158] In April 2011, the D.C. Circuit dismissed the challenge stating it did not have jurisdiction.[159]

In 2006, California passed the Global Warming Solutions Act (GWSA) of 2006, A.B. 32.[160] The GWSA requires a reduction of GHG emissions to 1990 levels (427 million metric tons) by the year 2020.[161] Environmental-justice advocates challenged CARB's plan to implement the law, claiming that the plan failed to minimize GHGs as required by the law and threatened vulnerable communities.[162] Among other things, the plaintiffs specifically objected to CARB's plan to use a cap-and-trade program to implement A.B. 32's mandate because CARB allegedly failed to conduct an adequate analysis of alternatives to cap-and-trade, which was required by California's Environ-

---

federal standards; (2) does not need such standards to meet compelling and extraordinary conditions; or (3) has proposed standards not consistent with section 202(a) of the Clean Air Act.

154. *See* Letter from Stephen Johnson, Adm'r, EPA, to Governor Arnold Schwarzenegger (Dec. 19, 2007), *available at* http://www.epa.gov/otaq/climate/20071219-slj.pdf.

155. Presidential Memorandum, State of California Request for Waiver Under 42 U.S.C. 7543(b), the Clean Air Act, 74 Fed. Reg. 7,040 (Feb. 12, 2009), *available at* http://www.whitehouse.gov/the-press-office/Presidential-Memorandum-EPA-Waiver (Jan. 26, 2009).

156. 74 Fed. Reg. 32,744 (July 8, 2009). President Obama's memorandum and EPA's subsequent decision to waive the Clean Air Act preemption in favor of California's motor-vehicle emissions standards are likely symptomatic of an increasing federal commitment to reducing GHG emissions.

157. *See Vehicle Greenhouse Gas Emissions Standards*, Ctr. for Climate & Energy Solutions, http://www.c2es.org/us-states-regions/policy-maps/vehicle-ghg-standards (last updated July 5, 2012).

158. *See* U.S. Chamber of Commerce v. EPA, No. 09-1237 (D.C. Cir. filed Sept. 9, 2009). California, New York, Arizona, Connecticut, Delaware, Florida, Illinois, Iowa, Maine, Maryland, Minnesota, Massachusetts, New Jersey, New Mexico, Oregon, Pennsylvania, Rhode Island, Vermont, Washington, and several environmental groups moved to intervene in the lawsuit.

159. Chamber of Commerce of the U.S. v. EPA, 642 F.3d 192, 203–06 (D.C. Cir. 2011).

160. Assemb. B. No. 32 (Cal. 2006).

161. Cal. Health & Safety Code § 38550.

162. Ass'n of Irritated Residents v. Cal. Air Res. Bd., No. CPF-09-509562 (Cal. App. Dep't Super. Ct. filed June 10, 2009).

mental Quality Act.[163] After the San Francisco Superior Court issued an injunction preventing CARB from conducting any further work on its cap-and-trade program, the California Court of Appeals stayed the injunction, thereby allowing CARB to continue its work on the cap-and-trade program.[164] CARB subsequently approved a revised analysis of alternatives to cap-and-trade to address the San Francisco Superior Court's original ruling and the plaintiffs' complaint.[165] In October 2011, CARB passed a cap-and-trade regulation that would allow for the first phase of enforcement to begin on January 1, 2013, a year later than originally planned, with a second phase to follow in 2015.[166]

Even with CARB's official adoption of the cap-and-trade program, judicial challenges continue to emerge and may impact California's overall plans to regulate GHGs. In March 2012, two environmental organizations challenged the cap-and-trade regulation by seeking to enjoin the state from allowing emission offset credits to be used for compliance with emissions-reduction requirements under the program.[167] Other litigation may soon emerge, as supporters and opponents of A.B. 32 debate how cap-and-trade revenue may be disbursed.[168]

California's recent budget woes may impact California's overall plans to regulate GHGs. Nevertheless, California likely will continue to lead the states in GHG regulation. Insurers should keep potential jurisdictional differences in mind as they prepare their underwriting analyses.

G. Regional Agreements

With leadership from governors, three regions of the United States have begun to develop GHG reduction agreements between states. Each of these regional cap-and-trade programs is at a different stage of development. The agreements allow GHG

---

163. *Id.*
164. *Id.*; Cal. Air Res. Bd. v. Ass'n of Irritated Residents, No. A132165 (Cal. Ct. App. June 24, 2011).
165. Cal. Air Res. Bd. Res. 11-27, AB 32 Scoping Plan and the Final Supplement to the AB 32 Scoping Plan Functional Equivalent Document (FED), *available at* http://www.arb.ca.gov/cc/scopingplan/final_res_scoping_plan_08242011.pdf.
166. News Release, Cal. Air Res. Bd., California Air Resources Board Adopts Key Element of State Climate Plan (Oct. 20, 2011), *available at* http://www.arb.ca.gov/newsrel/newsrelease.php?id=245.
167. Petition for Writ of Mandate & Complaint for Declaratory Injunctive Relief, Citizens Climate Lobby and Our Children's Earth Found. v. Cal. Air Res. Bd. (Cal. Super. Ct. S.F. County, filed March 27, 2012), *available at* http://www.peer.org/docs/epa/3_28_12_Cal_GHG_Complaint.pdf; Kevin Haroff, *Public Interest Groups First to Challenge California Cap-and-Trade Rules*, MARTEN LAW (Apr. 16, 2012), http://www.martenlaw.com/newsletter/20120416-calif-cap-and-trade-rules-challenge.
168. DALLAS BURTRAW & SARAH JO SZAMBELAN, A PRIMER ON THE USE OF ALLOWANCE VALUE CREATED UNDER CALIFORNIA'S CO_2 CAP-AND-TRADE PROGRAM 12–16 (2012), *available at* http://next10.org/sites/next10.org/files/20120504_Primer_Revised_V5.pdf.

regulation and carbon markets to begin in the United States while the federal government considers a national approach.

Regional Greenhouse Gas Initiative (RGGI). RGGI is the first mandatory, market-based effort in the United States to reduce GHG emissions. In 2005, governors of participating states signed a legally binding Memorandum of Understanding (MOU) committing their states to work cooperatively to reduce GHG emissions.[169] Under the MOU, states are responsible for implementation of these commitments within their individual state regulatory processes. Participating states included Connecticut, Delaware, Maine, Maryland, Massachusetts, New Hampshire, New Jersey, New York, Rhode Island, and Vermont. New Jersey, however, withdrew from the RGGI in May 2011.

In the MOU, participating states agreed to cap and reduce CO_2 emissions by 10 percent from the power sector by 2018.[170] The states' cap-and-trade approach includes requirements for fossil-fuel-fired electric power generators with a capacity of 25 megawatts or greater ("regulated sources") to hold allowances equal to their CO_2 emissions over a three-year control period; allocating CO_2 allowances through quarterly regional CO_2 allowance auctions; investing proceeds from the CO_2 allowance auctions in consumer benefit programs to improve energy efficiency and accelerate the deployment of renewable energy technologies; allowing offsets (GHG emissions reduction or carbon sequestration projects outside the electricity sector) to help companies meet their compliance obligations; and an emissions and allowance tracking system to record and track RGGI market and program data, including CO_2 emissions from regulated power plants and CO_2 allowance transactions among market participants.[171]

The regulatory authority of the individual states is the basis for the RGGI CO_2 Budget Trading Programs.[172] Through independent regulations based on a cooperatively designed Model Rule, each state limits emissions of CO_2 from electric power plants, creates CO_2 allowances, and establishes participation in CO_2 allowance auctions.[173] Regulated power plants can trade CO_2 allowances issued by any of the participating states to demonstrate compliance with an individual state program. CO_2 offset allowances also may be used to satisfy a limited portion of a regulated power plant's compliance obligation.[174] In this manner, the state programs, in aggregate, function as a single regional

---

169. Memorandum of Understanding on the Regional Greenhouse Gas Initiative From Governors, (Dec. 20, 2008), *available at* http://rggi.org/docs/mou_12_20_05.pdf.
170. *Id.* at 3.
171. *Program Overview*, RGGI, http://rggi.org/design/overview (last visited Aug. 20, 2012).
172. *State Statutes and Regulations*, RGGI, http://rggi.org/design/regulations (last visited Aug. 20, 2012).
173. *Id.*
174. RGGI limits the award of offset allowances to five project categories designed to mitigate or sequester CO_2, CH_4, or SF_6 within the ten-state region. *CO_2 Offsets*, RGGI, http://www.rggi.org/market/offsets (last visited Aug. 20, 2012).

market for CO_2 emissions. The first RGGI auction took place on September 25, 2008, and as of June 2012, RGGI had managed 15 subsequent auctions.[175]

Western Climate Initiative (WCI). The goal of the WCI was to develop a cap-and-trade program by January 1, 2012 and a multi-sector regulatory program by 2015.[176] Originally, the WCI participants included Arizona, British Columbia, California, Manitoba, Montana, New Mexico, Ontario, Oregon, Quebec, Utah, and Washington, but by the end of 2011 all of the U.S. states except California had dropped out.[177] Both California and Quebec adopted cap-and-trade systems in late 2011 that are likely to go into effect in 2013. California and Quebec also are planning to link their emissions trading systems.[178]

H. State Insurance Regulation

1. The National Association of Insurance Commissioners (NAIC)

Initially, the National Association of Insurance Commissioners (NAIC) formed its Climate Change and Global Warming (EX) Task Force (NAIC Task Force) to consider the relationship between climate change and insurance. The NAIC Task Force looked at investments and potential opportunities for the industry but mainly focused on disclosure of climate change-related risks. On May 31, 2008, the NAIC Task Force issued a white paper addressing various topics including disclosure, investment, and opportunities for the industry.[179] In the white paper, the NAIC Task Force explained its pursuit of mandatory climate change disclosures from insurers as an effort to (a) ensure the financial stability of insurers who may be both directly and indirectly impacted by climate change–driven weather-related risks and (b) impede reductions in coverage availability of personal and commercial property insurance, which is a predicted likely outcome of global warming.[180] The NAIC Task Force encouraged regulators to recognize the risk of weather-related losses on real estate that may, in part, back the investments used by insurers to fund their reserves.[181] The NAIC Task

---

175. *Auction Results*, RGGI, http://rggi.org/market/co2_auctions/results (last visited Aug. 20, 2012).
176. *Program Design*, WCI, http://www.westernclimateinitiative.org/designing-the-program.
177. *Id.*; Alessandro Vitelli & Michael B. Marois, *Six U.S. States Withdraw from Emissions-Trading Group Western Climate Pact*, Vancouver Sun (Nov. 17, 2011), http://www.vancouversun.com/business/states+withdraw+from+emissions+trading+group+Western+Climate+Pact/5726883/story.html#ixzz1khOdVevl.
178. Valerie Volcovici, California on track to link CO_2 scheme with Quebec in 2013, Reuters, Oct. 2, 2012, available at http://www.reuters.com/
179. Nat'l Ass'n of Ins. Comm'rs, Climate Change and Global Warming (EX) Task Force, The Potential Impact of Climate Change on Insurance Regulation (2008).
180. *Id.*
181. *Id.* at 1–2.

Force also noted the risk to insurers from investing in sectors of the economy that may have heavy exposure to global warming effects.[182]

The NAIC Task Force white paper also outlined potential opportunities that may arise out of climate change challenges.[183] For example, as new economic sectors emerge to provide goods and services that reduce GHG emissions or that are carbon neutral, there will be increased investment opportunities in the infrastructure needed to move to "cleaner" or more efficient energy sources.[184] Insurers may be able to hedge against potential claims from catastrophic events by investing in the commodities that will be in demand should such events occur.[185]

a. NAIC Climate Risk Disclosure Survey

After preparing the white paper, the NAIC Task Force continued work on preparing a climate risk disclosure survey for insurers. Insurance commissioners originally voted to make the survey mandatory in March 2009, but then reversed course in the spring of 2010 by voting to allow insurers to submit their responses on a voluntary, confidential basis.

In March 2009, the NAIC Task Force adopted a mandatory requirement that insurance companies disclose to regulators the financial risks they face from climate change as well as actions the companies are taking to respond to those risks.[186] According to the 2009 mandate, insurance companies with annual premiums in excess of $500 million were required to complete an Insurer Climate Risk Disclosure Survey each year.[187] Surveys were to be submitted to the domestic regulators of the insurer group's lead state (i.e., the regulator overseeing the insurer within the group that reports the largest direct written premium volume).[188] The first survey responses were set to be due May 1, 2010.

Both insurers and regulators raised concerns about the protection of proprietary information, the authority to request the information, and the logistics associated with administering a new survey. In response to such concerns, the NAIC Task Force, in a dramatic turn of events, voted at its spring 2010 meeting to no longer make the survey mandatory and public, but instead to allow insurers to submit their responses on a voluntary, confidential basis.[189] According to the revised plan, states that decide

---

182. *Id.*
183. *Id.* at 2.
184. *Id.*
185. *Id.*
186. *NAIC Insurer Climate Risk Disclosure Survey* (May 2009).
187. *Id.*
188. *Id.*
189. *Climate Risk Disclosure Survey: Adopted Version*, NAIC, (Mar. 28, 2010) http://www.naic.org/documents/committees_explen_climate_survey_032810.pdf.

to participate in the new voluntary, confidential survey will coordinate with NAIC to develop a public report providing information in the aggregate.[190]

The majority of state regulators determined not to administer the survey. The NAIC reports that only 21 states decided to administer the survey.[191] Of those states, only four—California, New York, Pennsylvania, and Washington—initially made the survey mandatory and the results public for the 2010 reporting year.[192] Connecticut, Maryland, Puerto Rico, and Vermont made participation mandatory but the full results private.[193] Florida, New Hampshire, New Jersey, and Oregon decided to make the results public, but made participation in the survey voluntary.[194]

b. 2011 Ceres Report on 2010 NAIC Survey Results

In September 2011, Ceres issued the report *Climate Risk Disclosure by Insurers: Evaluating Insurer Responses to the NAIC Climate Disclosure Survey*, analyzing what 88 U.S. insurers disclosed about climate change in the 2010 survey responses.[195] According to Ceres, the results show that the insurance industry's response to climate change is "sluggish" and "uneven," which could undermine the industry's own financial viability as well as the stability of the larger global economy.[196]

Although it appears from the survey results that there is a broad consensus among insurers that climate change will have an effect on extreme weather events, Ceres reports that the vast majority of insurers do not have any formal policy to deal with climate change.[197] Overall, the largest multiline insurers and reinsurers (those with more than $1 billion in annual premiums) were the most likely to have a climate change policy, while none of the 18 property and casualty companies reported any formal policy despite more than half believing that climate change poses a risk to their business.[198] With regard to risk from insured losses, Ceres reports that the survey responses are too general to determine whether insurers expect to experience greater or more volatile losses and what sort of effect such losses might have on pricing.[199] As

---

190. *Id.* Despite making the survey voluntary and confidential, the survey questions remain the same as those in the earlier 2009 version.

191. NAIC, Update on Climate Risk Disclosure Survey (presented at Oct. 17, 2010 meeting).

192. Sharlene Leurig, Ceres, Climate Risk Disclosure by Insurers: Evaluating Insurer Responses to the NAIC Climate Disclosure Survey 16 (2011), *available at* http://www.ceres.org/resources/reports/naic-climate-disclosure.

193. *Id.*

194. *Id.* The following states made participation in the survey voluntary and the responses confidential: Alabama, Colorado, Louisiana, Mississippi, Missouri, Nebraska, Ohio, Oklahoma, and South Carolina. *Id.*

195. Leurig, *supra* note 192.

196. *Id.* at 3.

197. *Id.* at 18. Only 11 of the 88 reporting companies stated that they had any formal climate change policy. *Id.*

198. *Id.* at 21–22.

199. *Id.* at 22–25.

to liability risk, Ceres reports that only eight insurers cite liability exposure in their discussion of climate risks, and those that did, cite mainly D&O exposure for failure by insureds to properly disclose climate change risks.[200] Ceres is concerned that no insurer cited liability risk relating to ongoing tort litigation, especially given the potential scale of liability as well as significant defense costs.[201]

With regard to disclosures on what financial effect climate change risks may have on insurers, Ceres found the results discouraging. According to Ceres, more than 40 percent of insurers who see their company as having climate risk exposure provide no information on the potential effects climate change may have on the company's pricing, capital adequacy, or reinsurance requirements.[202] Of those that provided information, most were concerned that the potential for extreme weather events may drive market prices up, undermining affordability of insurance and reinsurance products.[203] Despite these concerns, Ceres reports that only 18 percent of the 88 insurers are taking any steps to manage these risks.[204] The remaining 82 percent stated that their present diversification, reinsurance coverage, and annual contract terms are sufficient to manage climate change risk.[205] Ceres believes that without more specific and detailed responses, the existing disclosures regarding the financial effect of climate change provide little foresight into impending dislocations in the insurance market.[206]

Ceres is concerned that insurers are relying too heavily on third-party catastrophe risk models that only marginally integrate changing extreme weather when pricing insurance products.[207] While two of the three catastrophe modelers are integrating climate change into some of their models, their work to date has focused on hurricanes, excluding perils such as floods, windstorms, and fires.[208] Despite current modeling inadequacies, Ceres has found that most insurers believe the current models to be adequate, with only the largest insurers finding them inadequate.[209] Ceres is concerned that "without explicit education and dialogue between reinsurers, modelers, brokers, and primaries, the gulf between the most sophisticated insurers [who have the ability to create their own models] and the rest of the industry in terms of the

---

200. *Id.* at 25.
201. *Id.* at 26.
202. *Id.* at 27.
203. *Id.* at 28–30.
204. *Id.*
205. *Id.*
206. *Id.* at 33.
207. *Id.* at 34.
208. *Id.* at 35, 37.
209. *Id.*

capacity to anticipate nonlinear climate change trends will persist," putting consumers and the industry at risk.[210]

Although most insurers recognize that climate change will affect insureds, Ceres found that only 15 percent of insurers think their own investments have definite exposure to climate risk.[211] In addition, Ceres reports that most insurers have no plans to reduce their operational GHG emissions.[212] According to Ceres' review of the results, the majority of insurers are not undertaking any activities to help their customers better understand and manage climate change risk.[213] Given these findings, Ceres made several recommendations to regulators and requested that the NAIC implement *mandatory* annual disclosure requirements.[214]

2011 survey submissions were due in May 2012. Thus, as of this writing, Ceres has not issued a 2012 report on the 2011 results.

c. Recent NAIC Activities

Currently, the Climate Change and Global Warming Working Group is the NAIC group considering climate change-related issues.[215] This working group reports to the Financial Condition Committee and the Examination Oversight Task Force.[216] In 2012, the charges of the working group included (1) reviewing the enterprise risk management efforts by carriers and how they may be affected by climate change and global warming, (2) investigating and receiving information regarding the use of modeling by carriers and their reinsurers concerning climate change and global warming and its possible impact on investments, and (3) reviewing the impact of climate change and global warming on insurance investments through a presentation by interested parties.[217] As of the beginning of 2012, the Climate Change and Global Warming Working Group had several subgroups working on various initiatives. First, the Impact of Climate Change Exam Subgroup was working on proposals to amend the 2013 Financial Condition Examiners Handbook to address climate change issues. Second, the Climate Change Survey Subgroup was working on reviewing the substance of the questions in the NAIC Climate Risk Disclosure Survey and potentially recommending updates. Third, the Climate Change Survey Subgroup was evaluating participation by U.S. insurers in the United Nations Environment Programme

---

210. *Id.* at 38.
211. *Id.* at 39.
212. *Id.* at 45.
213. *Id.* at 47.
214. *Id.* at 50–51.
215. *See* http://www.naic.org/committees_e_climate.htm. The NAIC Task Force was dissolved and replaced by this working group.
216. *Id.*
217. *Id.*

Finance Initiative (UNEP FI) Global Survey: Advancing the Role of the Insurance Industry in Climate Change Adaptation. The survey and other methods of reporting may evolve in the coming years.

d. California, New York, and Washington Mandatory Survey

In February 2012, the California insurance commissioner issued a press release announcing that it would be joining forces with New York and Washington to administer an Insurer Climate Change Disclosure Survey independent of the NAIC process.[218] The three states plan to require all insurers that write more than $300 million in premiums nationwide and do business in one of the three states to respond to the survey regardless of where insurers are domiciled. This represents a large portion of insurers doing business in the United States. The three states will use the same questions developed by the NAIC in 2009. The survey results will be made public. The three states plan to use Ceres to review and synthesize the results. As of this writing, the initial results of the survey are not yet available.

2. California Green Insurance Act of 2010

On September 28, 2010, the California Green Insurance Act of 2010 was signed into law. The act took effect January 1, 2011. As enacted, the act amends Insurance Code sections 926.1, 926.2, and 12939 by extending until January 2015 the requirement that each insurer admitted in California provide information biennially to the insurance commissioner on all its community-development investments and community development infrastructure investments in California. In addition, the act extends what is considered community development investments and community development infrastructure developments to include "green investments."[219] The act also extends until January 2015 the requirement that the insurance commissioner post on the Insurance Department's website the aggregate insurer community-development

---

218. *See* Press Release, Cal. Dep't of Ins., Insurance Commissioner Dave Jones Announces Multi-State Effort on Climate Risk Disclosure Survey (Feb. 1, 2012), *available at* http://insurance.ca.gov/0400-news/0100-press-releases/2012/release009-12.cfm. *See also* NAIC, Climate Change and Global Warming Working Group, Agenda, Mar. 12, 2012 (Attachment 1—Climate Change and Global Warming Working Group Feb. 9, 2012 Conference Call Minutes).

219. "'Green investments' means investments that emphasize renewable energy projects, economic development, and affordable housing focused on infill sites so as to reduce the degree of automobile dependency and promote the use and reuse of existing urbanized lands supplied with infrastructure for the purpose of accommodating new growth and jobs. 'Green investments' also means investments that can help communities grow through new capital investment in the maintenance and rehabilitation of existing infrastructure so that the reuse and reinvention of city centers and existing transportation corridors and community space, including projects offering energy efficiency improvements and renewable energy generation, including, but not limited to, solar and wind power, mixed-use development, affordable housing opportunities, multimodal transportation systems, and transit-oriented development, can advance economic development, jobs, and housing." CAL. INS. CODE § 926.1(e).

investments and community development infrastructure investments, and requires the commissioner to identify specifically those insurers who make "green investments." Finally, the act provides tax incentives, in the form of tax credits, to insurers who invest in community development financial institutions (CDFIs).[220]

3. Pay-As-You-Drive Regulations

Other green insurance regulations of note include California's Pay-As-You-Drive (PAYD) or usage based insurance regulation.[221] The California PAYD program "is a cutting-edge program that will create financial incentives for California motorists to drive less, leading to lower-cost auto insurance, less air pollution and a reduced dependence on foreign oil."[222] The PAYD regulations allow drivers to purchase a more mileage-accurate insurance option.[223] The regulations were revised in the summer of 2009 and took effect in October 2009.[224] Insurance companies are offering PAYD discounts to drivers in California and numerous other states.

---

220. CDFIs are specialized financial institutions that are dedicated to (1) providing financial products and services to people and communities underserved by traditional financial markets and (2) supporting renewable energy projects, energy efficiency improvements, economic development, and affordable housing in these communities.

221. CAL. CODE REGS. tit. 10, § 2632.5 (2012).

222. Press Release, Cal. Dep't of Ins., Commissioner Poizner Releases New Draft of Pay-As-You-Drive Regulations (Aug. 3, 2009), *available at* http://www.insurance.ca.gov/0400-news/0100-press-releases/0080-2009/release117-09.cfm.

223. *Id.*

224. Cal. Dep't of Ins., Amendments to CAL. CODE REGS. tit. 10, § 2632.5, Final Regulation Text (Oct. 15, 2009), *available at* http://www.insurance.ca.gov/0250-insurers/0800-rate-filings/upload/PAYDFINALTXTFILED101609.pdf; Press Release, Cal. Dep't of Ins., Commissioner Poizner Releases New Draft of Pay-As-You-Drive Regulations (Aug. 3, 2009).

CHAPTER 6
Climate Change-Related Litigation

Various types of climate change-related litigation have begun to emerge in the United States. Plaintiffs have filed tort claims based on public nuisance, negligence, and civil conspiracy theories against entities that emit GHGs, such as utilities and chemical companies. In such cases, plaintiffs have sought injunctive relief or money damages. In one case, an insurer sought a declaratory judgment regarding whether there is a duty to defend an insured in one of the initial climate change-related tort actions. More recently, plaintiffs also have filed lawsuits based on the public-trust doctrine against the federal government and state governments. As discussed in Chapter 5, trade associations and other organizations have challenged EPA rulemakings related to GHGs. Public interest groups also have challenged governmental actions such as granting particular kinds of permits for facilities with GHG and other types of emissions. Although various types of litigation are reviewed, this chapter focuses primarily on the litigation that could lead to a potential increase in insurance claims.

First, in Section A, this chapter explores climate change-related tort litigation to date and provides commentary on the future of such litigation and additional types of tort theories that may emerge, such as products liability. Second, this chapter provides a brief overview of the first climate change-related insurance case, in Section B. Third, this chapter discusses the public trust litigation filed in 2011, in Section C. Next, this chapter provides an overview and analysis of other types of litigation that may emerge in the future. Section D discusses the potential for natural resource damage claims based on CERCLA. Section E addresses potential shareholder, SEC, or state actions related to misrepresentation, concealment, or mismanagement of climate change-related risk. Section F analyzes the types of claims against service providers that could arise out of efforts to address or manage climate change-related impacts such as (1) claims related to preparation of environmental impact statements and mitigation strategies pursuant to federal and state environmental policy acts; (2) claims related to preparation of carbon footprints; (3) disputes related to emerging carbon markets; (4) greenwashing litigation related to allegedly green products; and (5) green building litigation. Section G discusses potential EPA enforcement actions.

Policyholders faced with climate change-related claims may seek coverage pursuant to (1) first-party property, accident, health, and political-risk policies, (2) third-party D&O, CGL, environmental, and professional/E&O policies, and (3) first-/third-party policies such as owners protective professional indemnity (OPPI) insurance and contractors protective professional indemnity (CPPI) insurance. Even if these claims are unsuccessful, they may nonetheless be costly to defend. Thus, this chapter provides a thorough description of recent and potential climate change-related litigation. Chapter 7 addresses the implications these kinds of climate change-related claims have for specific types of coverage.

A. Common Law Tort Litigation

1. Climate Change-Related Tort Litigation: An Overview

Climate change-related tort litigation in the United States has emerged fairly quickly. Plaintiffs filed the first such case, *American Electric Power Co. v. Connecticut (AEP)*, in 2004.[1] Since that time, plaintiffs have filed four other major climate change-related tort cases,[2] and *AEP* has been decided by the Supreme Court.[3]

To some, this signals that the "[f]lood gates of U.S. courts are beginning to open to global warming litigation with profound implications for companies across a broad swath of industries."[4] The nature of future climate change-related litigation will depend, in part, on the successes and failures in these early cases. But climate change litigation is also likely to follow the trends of past litigation. See Chapter 3, discussing historical emerging risks. There will probably be periods where plaintiffs are emboldened to file new suits followed by phases in which plaintiffs may temporarily step back to develop new theories and modes of attack. From an insurance perspective, climate change-related tort cases have implications for commercial general liability and environmental liability business lines. See Chapter 7.A.3.b–c. Below is a summary and assessment of the tort litigation to date in this area.

---

1. Am. Elec. Power Co. v. Connecticut, 131 S. Ct. 2527 (2011).
2. Native Vill. of Kivalina v. ExxonMobil Corp., No. 09-17490, ___ F.3d ___, 2012 WL 4215921 (9th Cir. Sept. 21, 2012), *petition for rehearing filed* Oct. 4, 2012; Comer v. Nationwide Mut. Ins. Co., No. 1:05 CV 436 LTD RHW, 2006 WL 1066645, at *1 (S.D. Miss. Feb. 23, 2006), *rev'd sub nom.* Comer v. Murphy Oil USA, 585 F.3d 855 (5th Cir. 2009), *rev'd en banc*, 607 F.3d 1049 (5th Cir. 2010) (refiled action pending in Mississippi federal court); California v. Gen. Motors Corp., No. C06-05755 MJJ, 2007 WL 2726871 (N.D. Cal. Sept. 17, 2007) (voluntarily dismissed); Turner v. Murphy Oil USA, Inc., 582 F. Supp. 2d 797, 800 (E.D. La. 2008) (dismissed upon settlement).
3. 131 S. Ct. 2527.
4. James M. Davis & Noel C. Paul, *Managing the Risks of Global Warming: Avoiding the Mass Tort Template and Insurance Coverage Implications*, 17 COVERAGE 5, 1 (2007).

2. *American Electric Power Co. v. Connecticut*

On June 20, 2011, the Supreme Court of the United States decided its first climate change-related tort case, *AEP*.[5] The plaintiffs asserted federal and state common law nuisance claims against electric power companies burning fossil fuels. They sought injunctive relief in the form of emissions caps on stationary-source GHG emitters. The Supreme Court rejected plaintiffs' federal common law nuisance claim. The court held 8–0 that the Clean Air Act, and the EPA regulatory activity it authorizes, displaces federal common law in this area. But a divided court left open the possibility of viable state law actions, remanding the case for further consideration.

In *AEP*, several states and nonprofit land trusts[6] commenced a lawsuit in 2004 seeking an order requiring that defendant power utility companies abate the public nuisance of global warming. Plaintiffs alleged that defendants[7] are "substantial contributors to elevated levels of carbon dioxide and global warming," as their annual emissions compromise "approximately one quarter of the U.S. electric power sector's carbon dioxide emissions and approximately ten percent of all carbon dioxide emissions from human activities in the United States."[8] Plaintiffs claimed that by contributing to global warming, defendants were harming the environment, the states' economies, and public health. Plaintiffs did not seek damages. Instead, they sought the following remedies: (1) an order holding each of the defendants jointly and severally liable for creating, contributing to, and/or maintaining the public nuisance of global warming and (2) an order permanently limiting (enjoining) defendants' emissions and consequent contribution to global warming by requiring that defendants cap their emissions and then reduce them by a specified percentage each year.[9]

The trial court (the U.S. District Court for the Southern District of New York) dismissed plaintiffs' case on grounds that the lawsuit raised nonjusticiable political questions. The political question doctrine is primarily a function of the constitutional separation of powers.[10] "The political question doctrine excludes from judicial review those controversies which revolve around policy choices and value determinations constitutionally committed for resolution to the halls of Congress or the confines of the Executive Branch."[11] Under this doctrine, a court must dismiss an action where it

---

5. 131 S. Ct. 2527.
6. The plaintiff group was a coalition of states (Connecticut, New York, California, Iowa, New Jersey, Rhode Island, Vermont, Wisconsin), the City of New York, and three land trusts.
7. The defendants included AEP, American Electric Power Service Corp., Southern Company, Tennessee Valley Authority, Xcel Energy Inc., and Cinergy Corp.
8. Connecticut v. Am. Elec. Power Co., 582 F.3d 309, 316 (2d Cir. 2009).
9. *Id.* at 318.
10. Baker v. Carr, 369 U.S. 186, 210 (1962).
11. Japan Whaling Ass'n v. Am. Cetacean Soc'y, 478 U.S. 221, 230 (1986).

determines that the issues are too political and, thus, must be resolved by the legislative or executive branches of government. Defendants frequently have invoked this doctrine as a defense in the climate change context.[12] In this case, the district court reasoned that these climate-change issues were better suited for resolution by the political branches.[13]

The U.S. Court of Appeals for the Second Circuit disagreed. On September 21, 2009, a two-member panel of the circuit court vacated the trial court's dismissal and remanded the case for further proceedings.[14] The court explained that a decision by a single federal court regarding whether the emissions of six coal-fired power plants constitutes a public nuisance does not set a national or international emissions strategy and thus did not present a nonjusticiable political question.[15] The court then addressed and rejected the defendants' other reasons for blocking the plaintiffs' case. The court held the plaintiffs had standing to bring their claims because their current and future injuries (harm to the environment, harm to the states' economies, and harm to public health) were "fairly traceable" to and caused by defendants.[16] The court also held that plaintiffs could assert claims under the federal common law of nuisance.[17] And contrary to defendants' assertions, the court held that the Clean Air Act and its progeny did not displace plaintiffs' federal common-law public-nuisance claims.[18] Indeed, the court found no comprehensive federal GHG regulatory scheme warranting displacement.[19]

On March 5 and March 10, 2010, the Second Circuit denied petitions for rehearing or rehearing en banc,[20] leading defendants to petition the Supreme Court for a writ of certiorari on August 2, 2010.[21] In their petition, defendants asserted that the Second Circuit's decision "sets a precedent that threatens the basic operations of the

---

12. *See, e.g.,* Native Vill. of Kivalina v. ExxonMobil Corp., 663 F. Supp. 2d 863, 877, 880–83 (N.D. Cal. 2009), *aff'd on other grounds by* No. 09-17490, ___ F.3d ___, 2012 WL 4215921 (9th Cir. Sept. 21, 2012), *petition for rehearing* filed Oct. 4, 2012; Comer v. Murphy Oil USA, 585 F.3d 855, 860 (5th Cir. 2009), *rev'd en banc*, 607 F.3d 1049 (5th Cir. 2010) (refiled action pending in Mississippi federal court).

13. Connecticut v. Am. Elec. Power Co., 406 F. Supp. 2d 265, 271–74 (S.D.N.Y. 2005).

14. Connecticut v. Am. Elec. Power Co., 582 F.3d 309. Only two circuit judges (McLaughlin and Hall) participated in the written decision and order. The Honorable Sonia Sotomayor was originally assigned to the appeal and participated in oral argument but did not participate in the decision by reason of her August 2009 appointment to the U.S. Supreme Court.

15. *Id.* at 325–32.

16. *Id.* at 345–49.

17. *Id.* at 387–88. The Second Circuit did not address the state law public nuisance claims because the federal common law applied in the case.

18. *Id.* at 387–88.

19. *Id.*

20. Petition for Writ of Certiorari, Am. Elec. Power Co. v. Connecticut, 131 S. Ct. 813 (2011) (No. 10-174), 2010 WL 3054374, at *1.

21. *Id.*

broadest possible spectrum of the nation's businesses."[22] In addition, defendants asked for clarification as to a number of recurring questions, including "whether States and private plaintiffs have standing to seek, and whether federal common law provides authority for courts to impose, a *non-statutory*, judicially created regime for setting caps on greenhouse gas emissions based on 'vague and indeterminate nuisance concepts.'"[23]

On December 6, 2010, the Supreme Court agreed to hear the case.[24] The court agreed to review three issues: standing, displacement, and justiciability. With respect to the first issue, AEP and the other petitioners sought review of whether states and private parties have standing to seek a judicially fashioned cap on the GHG emissions on the five electric utility companies. Petitioners argued that the respondents failed to satisfy the standing requirement articulated in *Massachusetts v. EPA*.[25] Petitioners argued that the respondents' injuries were not fairly traceable to defendants' emissions, and would not be redressed by a favorable court decision. In response, the states argued that they had more firmly established Article III standing than the plaintiffs had in *Massachusetts v. EPA*. The acting solicitor general also argued on behalf of petitioner Tennessee Valley Authority that because the respondents' suit involved generalized grievances—which are more appropriately resolved by the representative branches—plaintiffs lacked "prudential standing."

The court was also asked to review whether a cause of action to cap emissions can be implied under federal common law, and whether such an action is displaced by the Clean Air Act. Respondents argued that a cause of action to limit carbon dioxide emissions can be implied under federal common law, despite the Clean Air Act's regulatory scheme addressing emissions. Respondents pointed out EPA had not yet regulated carbon dioxide emissions from stationary-source electric utilities, and thus the Clean Air Act and any related regulations do not displace their claims. Petitioners responded that Congress intended to displace such actions by enacting the Clean Air Act, which sought to abate sources of interstate pollution.

Finally, the court agreed to consider whether the case presented nonjusticiable political questions. Petitioners asked the court to address whether claims to cap the utilities' emissions at "'reasonable' levels ... would be governed by 'judicially discoverable and manageable standards' or could be resolved without 'initial policy determination[s] of a kind clearly for nonjudicial discretion.'"[26] Petitioners also argued that there were no judicially manageable standards for claims like this one. Respondents, however,

---

22. *Id.* at *33.
23. *Id.* at *2 (citing City of Milwaukee v. Illinois, 451 U.S. 304, 317 (1981)).
24. Connecticut v. Am. Elec. Power Co., 582 F.3d 309 (2d Cir. 2009), *cert. granted*, 131 S. Ct. 813 (2010).
25. Massachusetts v. EPA, 549 U.S. 497 (2007).
26. Petition for Writ of Certiorari, Am. Elec. Power Co. v. Connecticut, 131 S. Ct. 813 (2011) (No. 10-174), 2010 WL 3054374, at *1 (citing Baker v. Carr, 369 U.S. 186, 217 (1962)).

argued that the suit was more appropriately categorized as an ordinary tort suit and emphasized that federal courts are adept at applying ordinary tort law standards.

On June 20, 2011, the court held that the Clean Air Act and the EPA actions it authorizes categorically displace any federal common-law right to seek abatement of carbon dioxide emissions from coal-fired power plants.[27] Contrary to the Second Circuit's holding, the Supreme Court held that EPA need not actually exercise its regulatory authority (i.e., set emissions standards) in order to displace federal common law; merely occupying the field of emissions regulations was sufficient. The court noted that "[t]he plaintiffs argue, and the Second Circuit held, that federal common law is not displaced until EPA actually exercises its regulatory authority, i.e., until it sets standards governing emissions from the defendants' plants."[28] The court held that it was irrelevant that the Clean Air Act permits emissions until EPA acts.[29] "The critical point is that Congress delegated to EPA the decision whether and how to regulate carbon-dioxide emissions from power plants; the delegation is what displaces federal common law."[30] Thus, the court rejected plaintiffs' argument that the cause of action was not displaced.

The plaintiffs also could not use federal common law to upset the EPA's expert determination.[31] The court noted that plaintiffs should seek administrative and judicial recourse through the Clean Air Act.[32] If plaintiffs are dissatisfied with EPA's course of action, their recourse under federal law is to follow Clean Air Act procedures and seek court of appeals review.[33] As to the court's central displacement holding, Justice Samuel Alito, joined by Justice Clarence Thomas, stated that he concurred in the displacement holding on the assumption that the interpretation of the Clean Air Act in *Massachusetts v. EPA* was correct.

The energy industry had hoped the court would go even further and clearly limit its holding on the issue of standing in *Massachusetts v. EPA*.[34] In that case, the court held that the State of Massachusetts had standing to challenge EPA's refusal to regulate GHGs under the Clean Air Act.[35] Litigants had questioned whether this holding was based on Massachusetts's entitlement to "special solicitude" in the standing analysis because of its quasi-sovereign interests, or whether Massachusetts could have met the typical Article III standing test without any special treatment. *AEP* did not settle the issue.

---

27. Am. Elec. Power Co. v. Connecticut, 131 S. Ct. 2527 (2011).
28. *Id.* at 2538 (quoting Milwaukee v. Illinois, 451 U.S. 304, 324 (1981)).
29. *Id.* at 2538–39.
30. *Id.* at 2538.
31. *Id.* at 2539.
32. *Id.*
33. *Id.*
34. 549 U.S. 497 (2007).
35. *Id.*

In *AEP*, the court was split (4–4) on whether plaintiffs had standing to bring their claims.[36] The court noted that only four of eight justices would hold that at least some (but presumably not all) of the plaintiffs have Article III standing under *Massachusetts v. EPA*.[37] The result was an affirmance of the Second Circuit's determination that plaintiffs did have standing (and that the district court therefore had jurisdiction to hear the case), but a standing ruling that is not binding on other circuits. Importantly, the court's decision does not provide the jurisdictional bar to future climate change tort lawsuits that the energy industry wanted. Thus, the scope of the *Massachusetts v. EPA* holding will probably be the subject of further litigation.[38]

The *AEP* decision is important for several reasons. First and foremost, the court held that the Clean Air Act and the EPA regulatory activity that it authorizes categorically displaces any federal common law right to seek abatement of carbon dioxide emissions from coal-fired power plants. Congress's delegation of regulatory authority under the Clean Air Act to EPA, not the extent of EPA's specific regulatory actions on the issue, is what displaces the federal common law of nuisance.

Second, the lack of a precedential standing holding is significant because the utilities had sought to pare back the Supreme Court's standing holding in *Massachusetts v. EPA*. This has potentially broad implications for a range of environmental challenges.

Third, the court's holding was not based on other threshold issues (i.e., the political question doctrine or prudential standing), and the court barely mentioned those doctrines in its decision.

Fourth, the court's ruling may encourage further regulatory action by EPA or lawsuits filed against EPA. The court held that federal common law is displaced because EPA is the congressionally chosen expert on the subject of GHG emissions and that plaintiffs have a remedy under the Clean Air Act if they are not satisfied with EPA's regulations, decision not to regulate, or lack of progress in regulating GHGs. This may spur EPA to take further and swifter regulatory action with respect to GHGs.

Fifth, the case only addressed federal common law nuisance claims (as opposed to state tort claims); it addressed claims for injunctive relief, not for damages under state tort law. This leaves open the possibility that the plaintiffs' bar will contend that these distinctions make a difference.

---

36. *Am. Elec. Power Co.*, 131 S. Ct. at 2535.
37. *Id.* (citing Massachusetts v. EPA, 549 U.S. 497 (2007)).
38. *Id.* The court also noted that four justices would have rejected the power companies' "prudential standing" arguments. The power companies sought dismissal "because of a 'prudential' bar to the adjudication of generalized grievances, purportedly distinct from Article III's bar." *Id.* at 2535 n.6. The court's statements on these "other threshold issues" are limited, and the court's holding is not based on these doctrines. Thus, parties will probably raise these defenses again in subsequent climate change-related tort cases.

Finally, the court could not squarely address the state tort preemption issue. "[T]he availability *vel non* of a state lawsuit depends, *inter alia*, on the preemptive effect of the [Clean Air Act]."[39] The court sent that issue back to the lower court on remand, but in September 2011, plaintiffs withdrew their complaints. Thus, this will likely be a pivotal issue in future climate change-related tort litigation. This issue might eventually be evaluated in *Kivalina* or *Comer*. This issue will not be addressed on remand in *AEP* because in September 2011, plaintiffs decided to withdraw their state law claims.

3. *Kivalina v. ExxonMobil Corp.*

In *Kivalina v. ExxonMobil Corp.*, plaintiffs alleged that 20 oil, coal, and electric utility companies have emitted large quantities of CO_2 through their operations and that these emissions have caused the melting of Arctic sea ice.[40] This Arctic sea ice had formerly protected the village of Kivalina, Alaska, from winter storms.[41] Plaintiffs alleged that because of the ice melt, these storms have eroded the coastline such that houses and buildings are in imminent danger of falling into the sea.[42] Plaintiffs sought monetary damages for defendants' "past and ongoing contributions to global warming, a public nuisance, and damages caused by certain defendants' acts in furthering a conspiracy to suppress the awareness of the link between these emissions and global warming."[43]

In their specific causes of action against defendants, plaintiffs alleged that defendants' emissions constitute a public nuisance and that alleged erosion of the Kivalina coastline is a direct and proximate result of defendants' emissions.[44] Plaintiffs also asserted civil conspiracy and concert of action claims against defendants.[45] Plaintiffs alleged that defendants have agreed to intentionally create, contribute to, and/or maintain the public nuisance of global warming.[46] Moreover, plaintiffs alleged that defendants have conspired to "mislead the public with respect to the science of global warming and to delay public awareness of the issue."[47]

In response to plaintiffs' complaint, defendants filed multiple motions to dismiss. Defendants argued that the case should be dismissed because plaintiffs' claims present

---

39. *Id.* at 2540.
40. Complaint for Damages and Demand for Jury Trial ¶¶ 3–4, Native Vill. of Kivalina v. ExxonMobil Corp., No. 4:08-cv-01138-SBA (N.D. Cal. Feb. 26, 2008).
41. *Id.*
42. *Id.*
43. *Id.*
44. *Id.* ¶¶ 249–61.
45. *Id.* ¶¶ 268–82.
46. *Id.* ¶ 269.
47. *Id.*

nonjusticiable political questions better left to the executive and legislative branches.[48] In addition, defendants argued that plaintiffs' state and federal nuisance claims should be dismissed because plaintiffs cannot prove that defendants caused the harm suffered by plaintiffs.[49] Defendants also argued that plaintiffs have no standing to assert federal common law nuisance claims, and that in any event, the Clean Air Act displaced the authority of the federal courts to apply federal common law to plaintiffs' claims.[50] Finally, defendants asserted that any conspiracy claims or concert of action claims must fail because they are not independent torts, but instead are a means of assigning derivative liability for an underlying tortious act.[51] Thus, such secondary claims must fail because plaintiffs cannot prove the primary nuisance claims.[52]

On September 30, 2009, the district court dismissed plaintiffs' action.[53] The court held that the federal nuisance claim presented nonjusticiable political questions and could not meet the "fairly traceable" standard for causation for Article III standing. The court also dismissed the plaintiffs' state claims without prejudice based on the court's discretion not to decide pendent state law claims.[54]

With regard to the political question doctrine, the court found no workable standards for a jury to decide whether defendants' emissions caused more harm (erosion to the Kivalina coastline) than good (providing power, utilities, and oil to industry and residences).[55] The court acknowledged that neither the Constitution nor any federal law prescribes the issues in the case to a decision by the political branches.[56] But it nonetheless held that the political questions doctrine barred review. The court reasoned that the issues in the case—the allowable amount of GHGs defendants can

---

48. Notice of Motion and Motion of Certain Oil Company Defendants to Dismiss Plaintiffs' Complaint Pursuant to Fed. R. Civ. P. 12(b)(1), Memorandum Points and Authorities (Document No. 135) at 2, *Native Vill. of Kivalina v. ExxonMobil Corp.*, No. 4:08-cv-01138-SBA (N.D. Cal. June 30, 2008).

49. Notice of Motion and Motion of Certain Oil Company Defendants to Dismiss Plaintiffs' Complaint Pursuant to Fed. R. Civ. P. 12(b)(6), Memorandum Points and Authorities (Document No. 134) at 4, *Native Vill. of Kivalina v. ExxonMobil Corp.*, No. 4:08-cv-01138-SBA (N.D. Cal. June 30, 2008).

50. Motion of Certain Utility Defendants to Dismiss Plaintiffs' Civil Conspiracy Claim (Document No. 140) at 4, *Native Vill. of Kivalina*, No. 4:08-cv-01138-SBA; Notice of Motion and Motion to Dismiss of Defendant Peabody Energy Corp. for Lack of Subject Matter Jurisdiction Pursuant to Fed. R. Civ. P. 12(b)(1) and for Failure to State a Claim upon Which Relief May Be Granted Pursuant to Fed. R. Civ. P. 12(b)(6) (Document No. 137) at 9, *Native Vill. of Kivalina*, No. 4:08-cv-01138-SBA.

51. Motion to Dismiss of Defendant Peabody Energy (Document No. 137), *supra* note 50, at 22; Motion to Dismiss of Certain Oil Company Defendants (Document No. 134), *supra* note 49, at 20.

52. *Id.*

53. *Native Vill. of Kivalina v. ExxonMobil Corp.*, 663 F. Supp. 2d 863 (N.D. Cal. 2009), *aff'd on other grounds by* No. 09-17490, ___ F.3d ___, 2012 WL 4215921 (9th Cir. Sept. 21, 2012), *petition for rehearing* filed Oct. 4, 2012.

54. *Id.* at 877, 880–83.

55. *Id.* at 876–77.

56. *Id.* at 872–73.

emit and who should bear the cost of global warming—requires the court to make an initial policy determination that is best left to the political branches.[57]

Next, the court held that even if the case did not implicate political questions, plaintiffs do not have standing to bring their federal nuisance claim. The court cited four reasons the plaintiffs' allegations did not meet the low "fairly traceable" standard for causation for Article III standing.[58]

First, the court held that plaintiffs could not use a contribution theory of causation to satisfy the fairly traceable standard.[59] Plaintiffs' argument that the Clean Water Act (CWA) provides precedent for using the contribution theory fails.[60] The court reasoned that this case could not be analogized to CWA cases because the CWA provides federal limits on pollution in waterways, thus providing the assumption that any discharge over that amount is harmful.[61] Plaintiffs cannot make the same assumption here because the federal government has not chosen to regulate GHG emissions in that manner.[62]

Second, even if the contribution theory were applicable, the court held that plaintiffs do not allege sufficient facts to show that defendants' emissions are the "seed" of plaintiffs' injury.[63] The court explained that the plaintiffs could not trace the erosion of the Kivalina coastline to any defendant's emissions.[64] Plaintiffs' allegations

> make [ ] clear that there is no realistic possibility of tracing any particular alleged effect of global warming to any particular emissions by any specific person, entity, [or] group at any particular point in time. Plaintiffs essentially concede that the genesis of global warming is attributable to numerous entities which individually and cumulatively over the span of centuries created the effects they now are experiencing.[65]

Third, the court held that any argument by plaintiffs that they are within the "zone of discharge" of defendants' emissions, thus negating the need to trace emissions directly to defendants, also fails because the link between the injuries alleged and the defendants' specific emissions is too tenuous.[66] If the court adopted plaintiffs' argument, the entire world would be within the "zone of discharge."[67]

---

57. *Id.*
58. The causation standard for Article III standing is a lower standard than the proximate cause standard for torts.
59. *Native Vill. of Kivalina*, 663 F. Supp. 2d at 879–80.
60. *Id.*
61. *Id.*
62. *Id.* at 880.
63. *Id.* at 880–81.
64. *Id.* at 880.
65. *Id.*
66. *Id.* at 881.
67. *Id.*

Finally, the court held that plaintiffs are not entitled to a relaxed standing requirement under the theory of "special solicitude" as applied to states in *Massachusetts v. EPA*.[68] Consequently, the court held that plaintiffs do not have standing to bring their federal nuisance claim and the claim must be dismissed.[69]

On November 5, 2009, the *Kivalina* plaintiffs appealed the case to the Ninth Circuit. The plaintiffs believe they can distinguish their federal nuisance claims from those asserted, and rejected, in *AEP*.[70] But the Ninth Circuit is not obligated to follow the Supreme Court's decision that the plaintiffs had standing to bring those claims because a majority of the court did not support that holding. Thus, the Ninth Circuit will address federal common law nuisance claims and related defenses such as the political question doctrine and standing.

As in *AEP*, the plaintiffs have state law claims, not just the federal common law of nuisance cause of action. The state causes of action are not at issue in the current Ninth Circuit appeal. Thus, the Ninth Circuit appeal will not dispose of all of the *Kivalina* plaintiffs' claims. The *Kivalina* plaintiffs therefore may file another action in state court based on state law.

Although it will be difficult for plaintiffs to prove that the alleged erosion was caused by climate change and that defendants substantially contributed to that climate change, California is likely a friendlier jurisdiction than some for such claims. In particular, California courts have recognized a "market share" theory of liability that plaintiffs have subsequently attempted to use in both asbestos and lead paint cases, but with very limited success.[71] See discussion of the market share liability theory in Section A.6.b.

On September 21, 2012, the Ninth Circuit issued its decision in *Native Village of Kivalina v. ExxonMobil Corporation*.[72] In this long-awaited opinion, the court held that the doctrine of federal displacement bars the *Kivalina* plaintiffs' federal common law public nuisance claim for damages. The Ninth Circuit decision applies and extends the Supreme Court decision in *AEP v. Connecticut*.[73] *AEP* held that the federal Clean Air Act displaces climate change-related federal common law public nuisance claims for injunctive relief. Under the doctrine of federal displacement, a federal common law claim is displaced (i.e., barred) if Congress, through a statute,

---

68. *Id.* at 882. The district court incorrectly names the "special solicitude" doctrine "special solitude."
69. *Id.* at 883.
70. Lawrence Hurley, *Impact of Supreme Court's Greenhouse Gas Ruling Likely to Be Felt in Other Cases*, N.Y. Times, June 21, 2011, *available at* http://www.nytimes.com/gwire/2011/06/21/21greenwire-impact-of-supreme-courts-greenhouse-gas-ruling-41463.html.
71. *See* Gramling v. Mallett, 701 N.W.2d 523, 586–87 (Wis. 2005) (discussing fungibility of asbestos and possible limited applicability to asbestos brake pads); Rhode Island v. Lead Indus. Ass'n, Inc., 951 A.2d 428 (R.I. 2008) (reversing jury verdict finding against lead paint manufacturers and the Lead Industry Association based on public nuisance and applying market share liability).
72. Native Vill. of Kivalina v. ExxonMobil Corp., No. 09–17490, ___ F.3d ___, 2012 WL 4215921 (9th Cir. Sept. 21, 2012), *petition for rehearing* filed Oct. 4, 2012.
73. 31 S. Ct. 2527 (2011).

has addressed the subject encompassed by the claim (e.g., greenhouse gas emission abatement). The *Kivalina* plaintiffs had tried to argue that *AEP* was distinguishable because it involved injunctive relief as opposed to a request for damages. According to the principal *Kivalina* opinion, the Supreme Court made it clear that "displacement of a federal common law right of action means displacement of remedies."[74] "Thus, *AEP* extinguished *Kivalina*'s federal common law public nuisance damage action, along with the federal common law public nuisance abatement actions."[75]

The opinion was authored by Circuit Judge Sidney Thomas and joined by Circuit Judge Richard Clifton and District Judge Philip Pro of the District of Nevada (sitting by designation). It was based on the federal displacement doctrine but not the political question doctrine or standing. Judge Pro wrote a concurring opinion that not only elaborates on the doctrine of federal displacement but also agrees with the district court that the plaintiffs lack Article III standing. According to Judge Pro, standing is lacking because "Kivalina has not met the burden of alleging facts showing Kivalina plausibly can trace their injuries to Appellees."[76] Instead, Judge Pro notes that "[b]y Kivalina's own factual allegations, global warming has been occurring for hundreds of years and is the result of a vast multitude of emitters worldwide."[77]

In his concurring opinion, Judge Pro also indicates that the plaintiffs are not necessarily without a judicial remedy. "Once federal common law is displaced, state nuisance law becomes an available option to the extent it is not preempted by federal law."[78] *AEP* and *Kivalina* did not specifically address the preemption question. The *Kivalina* plaintiffs sought rehearing en banc on October 4, 2012. They also may refile their state law nuisance claims in state court.

4. California v. General Motors Corp.

In an earlier but also recent climate change-related nuisance case, *California v. General Motors Corp.*, the State of California brought suit against six major motor vehicle manufacturers, General Motors, Ford, Toyota, Honda, Chrysler, and Nissan.[79] The complaint alleged a host of injuries to California, its environment, its economy, and the health and well-being of its citizens caused by defendants' production of millions of automobiles that collectively, and in mass quantities, released CO_2 in the atmosphere and contributed to an elevated level of CO_2 in the environment.[80]

---

74. *Native Vill. of Kivalina*, 2012 WL 4215921, at *5.
75. *Id.*
76. *Id.* at 16.
77. *Id.*
78. *Id.* at 14.
79. California v. Gen. Motors Corp., No. C06-05755 MJJ, 2007 WL 2726871 (N.D. Cal. Sept. 17, 2007).
80. *Id.* at *1.

The State of California alleged that defendants' vehicles emit approximately 289 million metric tons of CO_2 and other GHGs each year and that such emissions contribute to the public nuisance of global warming.[81] California alleged the following injuries: reductions in snow pack, coastal and beach erosion, seawater intrusion into freshwater areas, prolonged heat waves, and increased risk and intensity of wildfires.[82] California not only sought monetary damages for past and ongoing contributions to global warming but also a declaratory judgment that defendants are liable for future monetary damages.[83]

Defendant automakers moved to dismiss the complaint on various grounds, including that the case raised nonjusticiable issues properly reserved for resolution by the political branches (the political question doctrine), the complaint failed to state a valid nuisance claim under federal common law, the complaint failed to state a valid nuisance claim under California law, and federal law preempted the nuisance claim under California law.[84]

The court granted defendants' motion to dismiss on the grounds that the complaint raised nonjusticiable political questions that were beyond the federal court's jurisdiction. The court stated that "[j]ust as in [the trial court decision in] *AEP*, the adjudication of plaintiff's claim would require the Court to balance the competing interests of reducing global warming emissions and the interests of advancing and preserving economic and industrial development. The balancing of those competing interests is the type of initial policy determination to be made by the political branches, and not this Court."[85]

California initially appealed the decision to the U.S. Court of Appeals for the Ninth Circuit. On June 19, 2009, the California Attorney General's Office voluntarily dropped the appeal citing the following reasons for its dismissal: (1) EPA's April 17, 2009, acknowledgment that GHGs threaten the public health and welfare; (2) President Obama's announcement of a national program to reduce GHGs and improve fuel economy; and (3) defendants General Motors' and Chrysler's filings under Chapter 11 of the Bankruptcy Code.[86]

5. Tort Litigation Related to Episodic Climatic Events

a. *Turner v. Murphy Oil USA, Inc.*

In *Turner v. Murphy Oil*, plaintiffs, homeowners and business owners affected by a massive oil spill from the Murphy Oil refinery in Louisiana, sued Murphy Oil alleging that the spill occurred due to Murphy Oil's failure to properly secure and maintain

---

81. *Id.*
82. *Id.*
83. *Id.* at *2.
84. *Id.*
85. *Id.* at *8 (internal citation omitted).
86. Unopposed Motion to Dismiss Appeal at *2–3, Comer v. Nationwide Mut. Ins. Co., No. 06-cv-05755 MJJ (9th Cir. June 19, 2009).

its aboveground oil storage tanks to withstand the effects of Hurricane Katrina.[87] The federal district court certified the class and adopted a trial plan bifurcating the trial into two different phases—one addressing common issues of liability and general causation and the second consisting of successive trials on specific causation and compensatory damages.[88] The second phase would only take place if a jury found Murphy liable in whole or in part in the first phase. Prior to the the first phase, the parties agreed to settle the suit for $330,126,000, plus all reasonable fees and costs to be determined by the court.[89] Successful settlements such as this one are likely to inspire plaintiffs to bring more tort cases related to episodic climatic events.

> The plaintiffs' bar has a large war chest from the $330 million settlement in *Turner*.

b. *Comer v. Murphy Oil USA, Inc.*

In *Comer*, Gulf Coast property owners filed a series of allegations against various mortgage companies, insurance companies, oil companies, chemical manufacturers, and coal companies related to harm suffered from the effects of Hurricane Katrina.[90] First, in September 2005, plaintiffs filed a class action lawsuit against a number of insurance and mortgage companies.[91] The plaintiffs alleged that the insurers wrongfully denied coverage for hurricane-related damage pursuant to water damage exclusions and that the mortgage companies failed to obtain adequate insurance for the mortgaged properties.[92] In January 2006, plaintiffs moved to file a second amended complaint to add as defendants chemical manufacturers and oil companies alleged to have caused damage to plaintiffs' properties through their operations. Plaintiffs alleged that the emissions from these companies have contributed to global warming, which in turn contributed to increased sea levels and the ferocity of Hurricane Katrina.[93] Plaintiffs also requested that the court certify a class action suit against all defendants—insurance companies, mortgage companies, oil companies, and chemical manufacturers.[94]

---

87. Turner v. Murphy Oil USA, Inc., 234 F.R.D. 597, 602–03 (E.D. La. 2006).
88. *See* Turner v. Murphy Oil USA, Inc., 582 F. Supp. 2d 797, 800 (E.D. La. 2008).
89. *See* Turner v. Murphy Oil USA, Inc., 472 F. Supp. 2d 830, 845 (E.D. La. 2007).
90. Comer v. Nationwide Mut. Ins. Co., No. 1:05 CV 436 LTD RHW, 2006 WL 106645 (S.D. Miss. Feb. 23, 2006), *rev'd sub nom.* Comer v. Murphy Oil USA, 585 F.3d 855 (5th Cir. 2009), *rev'd en banc*, 607 F.3d 1049 (5th Cir. 2010).
91. *Id.* at *1.
92. *Id.*
93. *See id.*
94. *See id.*

The district court denied plaintiffs' request to file their second amended complaint because the claims were too disparate to proceed in one action.[95] The court held that the claims against the mortgage companies and insurers would depend on the particulars of the individual contracts and that the claims against the oil and chemical companies were tort not contract based. Thus, the court refused to certify the class.[96] With respect to the insurance and mortgage companies, the court ruled that each individual plaintiff would have to file an individual action against its respective mortgage and insurance companies.[97]

With regard to the claims against the oil and chemical companies, the court allowed plaintiffs to file a third amended complaint clarifying the claims against these defendants and adding any additional defendants.[98] Although the court allowed plaintiffs to continue their action against the oil and chemical companies, it observed "that there exists a sharp difference of opinion in the scientific community concerning the causes of global warming."[99] The court foresaw

> daunting evidentiary problems for anyone who undertakes to prove, by a preponderance of the evidence, the degree to which global warming is caused by the emission of greenhouse gasses; the degree to which the actions of any individual oil company, any individual chemical company, or the collective action of these corporations contribute, through the emission of greenhouse gasses, to global warming.[100]

Since the 2006 decision in *Comer*, the IPCC has issued its Fourth Assessment Report, taking a very strong position that most of the observed increase in climate change since the mid-20th century is very likely due to an increase in anthropogenic GHG concentrations.[101] The evolving state of the science could have impacts on future litigation.

Shortly after this *Comer* decision in April 2006, plaintiffs filed a third amended complaint asserting various tort theories, including nuisance, trespass, and civil conspiracy, against oil companies, coal companies, and chemical manufacturers.[102] As did the plaintiffs in *Kivalina*, the *Comer* plaintiffs alleged that oil companies, coal companies, and chemical manufacturers "knowingly and willfully" engaged in activities that

---

95. See id. at *2–3.
96. See id.
97. See id. at *2.
98. See id. at *4.
99. Id.
100. Id.
101. IPCC, Climate Change 2007—The Physical Science Basis 665–66, 728–29 (S. Solomon, D. Qin, M. Manning, Z. Chen, M. Marquis, K.B. Averyt, M. Tignor & H.L. Miller eds., 2007), *available at* http://www.ipcc.ch/pdf/assessment-report/ar4/wg1/ar4-wg1-chapter9.pdf.
102. Third Am. Class Action Compl., ECF No. 79, *Comer v. Murphy Oil, USA*, No. 1:05-CV-436-LG-RHW (S.D. Miss. Apr. 19, 2006).

contributed to global warming.[103] Plaintiffs' claims were predicated upon the "demonstrable changes" in the Earth's climate as a result of the defendants' GHG emissions.[104] The coal companies moved to dismiss the action on grounds that the plaintiffs lacked standing and that the claims were barred by the political question doctrine.[105] On August 30, 2007, the district court granted the motions to dismiss on those grounds.[106] At the same time, the district court *sua sponte* dismissed the remaining claims against all other defendants for plaintiffs' lack of standing.[107] Plaintiffs appealed to the Fifth Circuit on September 17, 2007.

On October 16, 2009, in a long-awaited decision, the Fifth Circuit partially reversed the district court and remanded the case for further proceedings.[108] First, the court determined that plaintiffs had standing to bring their nuisance, trespass, and negligence claims.[109] The defendants had argued that plaintiffs lacked standing because (1) the causal link between emissions, sea level rise, and Hurricane Katrina is too attenuated; and (2) the defendants' actions are only one of many contributions to GHG emissions.[110] The court noted that the plaintiffs' complaint, relying on scientific reports, alleged a chain of causation between defendants' substantial emissions and plaintiffs' injuries, and such allegations must be taken as true at that stage in the litigation.[111] Moreover, the court held that defendants' contentions were similar to those recently rejected by *Massachusetts v. EPA*.[112] In holding that Massachusetts had standing, the Supreme Court accepted as plausible the link between man-made GHGs and global warming.[113] The Fifth Circuit also noted that *Massachusetts v. EPA* held that injuries may be fairly traceable to actions that *contribute* to, rather than solely or materially cause, GHG emissions and global warming.[114] Based on *Massachusetts* and other cases, the Fifth Circuit concluded that plaintiffs satisfied the "fairly traceable" element of standing.[115]

---

103. Third Am. Class Action Compl. ¶ 11, *supra* note 102.
104. Third Am. Class Action Compl. ¶ 5, *supra* note 102.
105. Mem. of Authorities in Supp. of Coal Cos.' Mot. to Dismiss for Lack of Subject Matter Jurisdiction & for Failure to State a Claim at 5-10, ECF No. 147, Comer v. Murphy Oil, USA, No. 1:05-CV-436-LG-RHW (S.D. Miss. June 30, 2006).
106. Order Granting Defs.' Mot. to Dismiss, ECF No. 368, Comer v. Murphy Oil, USA, No. 1:05-CV-436-LG-RHW (S.D. Miss. Aug. 30, 2007).
107. Order Granting Defs.' Mot. to Dismiss, *supra* note 106.
108. Comer v. Murphy Oil USA, 585 F.3d 855 (5th Cir. 2009), *rev'd en banc*, 607 F.3d 1049 (5th Cir. 2010).
109. *Id.* at 867.
110. *Id.* at 865.
111. *Id.*
112. *Id.* at 865 (citing Massachusetts v. EPA, 549 U.S. 497 (2007)).
113. 549 U.S. at 523.
114. *Comer*, 585 F.3d at 866 (citing *Massachusetts*, 549 U.S. at 523).
115. *Id.* at 866.

Defendants argued that *Massachusetts* is distinguishable because Massachusetts received special consideration in the standing analysis as a state. The Fifth Circuit held that plaintiffs did not need "special solicitude" in *Comer* to get standing because the causal chain in *Comer* is "one step shorter" than in *Massachusetts*.[116]

Second, the Fifth Circuit held that plaintiffs' nuisance, trespass, and negligence claims did not present nonjusticiable political questions.[117] According to the Fifth Circuit, "[c]ommon law tort claims are rarely thought to present nonjusticiable political questions."[118] Because no constitutional or federal law provision specifically delegated the issues in these claims to a political branch, the court held that the issues do not present political questions and no further inquiry was required.[119]

Third, the court held that plaintiffs do not have standing to bring their unjust enrichment, fraudulent misrepresentation, and civil conspiracy claims.[120] These claims do not satisfy prudential standing requirements but represent "a generalized grievance that is more properly dealt with by the representative branches and common to all consumers of petrochemicals and the American public."[121]

As expected, defendants challenged the Fifth Circuit's decision in the form of a request for rehearing en banc.[122] After conducting a poll, a majority of the judges on the court granted the request.[123] After the en banc court had been impaneled, however, one of the nine judges recused himself causing the court to lose its quorum.[124] As a result, the court was no longer "authorized to transact judicial business" and, after considering available alternatives, the court dismissed the appeal, effectively reinstating the district court's decision.[125] On August 26, 2010, plaintiffs filed a petition for a writ of mandamus with the Supreme Court, requesting that the court either reinstate the

---

116. *Id.* at 865 n.5. The court further stated that:

> [u]nlike in other cases in which courts have held that traceability was lacking, the causal chain alleged by these plaintiffs does not depend on independent superseding actions by parties that are not before the court. Nor does it rely on speculation about what the effects of the defendants' actions will be, or about the actions anyone will take in the future. It involves only the predictable effects of the defendants' past actions on the natural environment, and the resulting harm to the plaintiffs' property.

Id. (internal citations omitted).
117. *Id.* at 869.
118. *Id.* at 873.
119. *Id.* at 875.
120. *Id.* at 867–68.
121. *Id.* at 868.
122. Comer v. Murphy Oil USA, 598 F.3d 208 (5th Cir.), *rev'd en banc*, 607 F.3d 1049 (5th Cir. 2010).
123. *Id.* at 210.
124. Comer v. Murphy Oil USA, 607 F.3d 1049, 1054 (5th Cir. 2010).
125. *Id.* at 1055.

Fifth Circuit's decision or suspend the Fifth Circuit's dismissal of plaintiffs' appeal.[126] The Supreme Court denied plaintiffs' petition on January 10, 2011.[127]

Ned Comer and fellow plaintiffs refiled their climate change tort action in the U.S. District Court for the Southern District of Mississippi on May 27, 2011.[128] The new action, based on diversity jurisdiction, alleges public and private nuisance, trespass, and negligence causes of action under Mississippi law. The complaint claims that plaintiffs suffered damages in Hurricane Katrina as a result of the production, exploration, mining, or combustion activities by the coal, oil, and chemical companies. As in the prior action, they argue that the defendants' GHG emissions made Hurricane Katrina more ferocious and damaging.

After all the procedural peculiarities in this case, the case was refiled. To continue their fight, plaintiffs relied on the following Mississippi statutory provision as a basis for refiling some of the same claims:

> If in any action, duly commenced within the time allowed, the writ shall be abated, or the action otherwise avoided or defeated, by the death of any party thereto, or for any matter of form, or if, after verdict for the plaintiff, the judgment shall be arrested, or if a judgment for the plaintiff shall be reversed on appeal, the plaintiff may commence a new action for the same cause, at any time within one year after the abatement or other determination of the original suit, or after reversal of the judgment therein, and his executor or administrator may, in case of the plaintiff's death, commence such new action, within the said one year.[129]

Plaintiffs apparently wanted this action on file when *AEP* was decided or simply wanted to file before the one-year deadline in this statute.

In March 2012, the Mississippi federal court granted the defendants' motions to dismiss and held that all of the plaintiffs' claims are barred by the doctrines of res judicata and collateral estoppel.[130] Alternatively, the court held that the plaintiffs do not have standing to assert their claims because their alleged injuries are not fairly traceable to the defendants' conduct.[131] Moreover, the court found that the lawsuit presented a nonjusticiable political question, and that all of the plaintiffs' claims were preempted by the

---

126. Petition for Writ of Mandamus, *In re Comer*, No. 10-294, 2010 WL 3493195, at *24–25 (U.S. Aug. 26, 2010).
127. *In re Comer*, 131 S. Ct. 902 (2011) (denying petition for writ of mandamus).
128. Class Action Complaint, Comer v. Murphy Oil USA, Inc., No. 1:11-cv-00220-LG-RHW (S.D. Miss. May 27, 2011).
129. Miss. Code Ann. § 15-1-69 (2012).
130. Memorandum Opinion and Order Granting Defendants' Motions to Dismiss, ECF No. 291, Comer v. Murphy Oil USA, Inc., No. 11-cv-00220-LG-RHW (S.D. Miss. Mar. 20, 2012).
131. Mem. Opinion & Order Granting Defs.' Mots. to Dismiss, *supra* note 130.

Clean Air Act.[132] The court further found that the plaintiffs' claims were barred by the applicable statute of limitations, and that "the plaintiffs cannot possibly demonstrate" that their injuries were proximately caused by the defendants' conduct.[133] On April 16, 2012, the plaintiffs filed a notice of appeal to the Fifth Circuit. A decision in 2013 is likely.

6. Future Trends: Climate Change-Related Tort Litigation

Businesses hit with climate change-related tort suits will look to the insurance industry to defend and indemnify them pursuant to current environmental liability or commercial general-liability policies or historic commercial general-liability policies, or through newly developed products to address the emerging market need. At first glance the exposure might not seem potentially significant. Indeed, plaintiffs asserting climate change-related common law claims face a number of potential hurdles including the political question doctrine defense, establishing standing, proving causation, and displacement or preemption by federal laws and regulations. But, transactional costs could still be high. Insurers could face responsibility for defense costs even if the initial set of claims is not successful.

a. Additional Types of Claims Including Products Liability

Plaintiffs' lawyers will continue to develop new climate change-related litigation theories. For example, in an interview published June 20, 2009, Gerald Maples, the lead plaintiff's attorney in *Comer*, predicted "massive litigation" in the future from "big farming interests" who suffer droughts, to "communities . . . ravaged by wildfires," to "ski resorts that have no snow."[134] Tort claims may come in many shapes and forms in future years.

A few commentators have theorized that individuals could bring products liability claims such as failure to warn or design defect claims related to climate change.[135] Such claims would allege that manufacturers failed to warn consumers that products emitted harmful GHGs or that products emitting GHGs that cause climate change are defectively designed.[136] Such claims would suffer from the same causation issues as other tort claims, but defense costs could nonetheless potentially be significant.

---

132. *Id.*

133. *Id.*

134. Paddy Manning, *You Are at Risk*, SYDNEY MORNING HERALD, June 20, 2009, *available at* http://www.smh.com.au/business/you-are-at-risk-20090620-crk4.html.

135. David A. Grossman, *Warming Up to a Not-So-Radical Idea: Tort-Based Climate Change Litigation*, 28 COLUM. J. ENVTL. L. 1, 39–52 (2003); Christina Ross, Evan Mills & Sean B. Hecht, *Limiting Liability in the Greenhouse: Insurance Risk-Management Strategies in the Context of Global Climate Change*, 26A STAN. ENVTL. L.J. 251, 292–93 (2007).

136. *Id.*

Thus, these types of claims should be considered in risk profiling for products-liability endorsements to CGL policies. See Chapter 7.A.3.b.

b. The Causation Issue

A market share theory of liability may provide plaintiffs with the ability to overcome causation issues in climate change litigation.

The plaintiffs' bar also will continue to develop new ways around the current roadblocks to success. Of particular importance, plaintiffs will attempt to overcome the difficulty of proving causation in climate change-related tort claims by (1) utilizing the evolving science of "attribution" and (2) advocating for the application of market-share liability theory principles, which effectively shift the burden of proof for causation to a defendant. At least one commentator has suggested that the foreseeability of emerging impacts related to climate change (e.g., sea level rise) based on available scientific information may provide an additional basis for holding governmental or private parties accountable for the effects of climate change.

(1) Attribution

As noted in Chapter 2, "attribution" is the science of trying to link particular climatic events to increases in GHG emissions. Scientists are trying to find "fingerprints" of GHG emissions in extreme weather events, such as the U.K. floods in 2000, European heat wave in 2003, and Moscow heat wave in 2010.[137] As the science of attribution continues to develop, plaintiffs may invoke such new theories in attempts to assign blame for damages to particular entities.

(2) Market Share Theory

The market share liability theory is a court-created doctrine for apportioning blame in the products liability context among an identified number of producers.[138] California

---

137. *See* John Carey, *Storm Warnings: Extreme Weather Is a Product of Climate Change*, Sci. Am., June 28, 2011 (discussing the use of the nascent science of attribution); Peter A. Stott et al., Attribution of Weather and Climate-Related Extreme Events (World Climate Research Programme, Climate Change in Service to Society Conference, Position Paper, Oct. 27, 2011), *available at* http://conference2011.wcrp-climate.org/documents/Stott.pdf (discussing recent work in the field of climate change attribution).

138. Generally, the market share theory requires:

(1) injury or illness occasioned by a fungible product (identical-type product) made by all of the defendants joined in the lawsuit; (2) injury or illness due to a design hazard, with each having been found to have sold the same type of product in a manner that made it unreasonably dangerous; (3) inability to identify the specific manufacturer of the product or products that brought about the plaintiff's injury or illness; and (4) joinder of enough of the manufacturers of the fungible or identical product to represent a substantial share of the market.

W. Page Keeton et al., Prosser and Keeton on the Law of Torts § 103, at 714 (5th ed. 1984).

and other states have used this theory to address causation in pharmaceutical products liability cases. The theory to apportion damages has been used in settlements in which a defendant's individual contribution to an alleged harm was difficult to determine. For example, as part of the 1998 tobacco settlement, tobacco industries agreed to pay damages to the 46 plaintiff states based on each company's market share in each state.[139]

The market share liability theory was initially adopted by the California Supreme Court in the 1980 case *Sindell v. Abbott Laboratories*.[140] In that case, a plaintiff brought a class action suit against a group of pharmaceutical companies alleging that the companies negligently manufactured, marketed, and promoted a drug administered to pregnant women to prevent miscarriages.[141] According to the plaintiff, defendants either knew or should have known that diethylstilbestrol (DES) was a carcinogen that could "cause cancerous and precancerous growths in the daughters" of women taking the drug.[142] A trial court dismissed the claim because the plaintiff could not identify which defendant was responsible for manufacturing the precise drug taken by her mother.[143] Applying the market share liability theory, the California Supreme Court reversed.[144] Although the case did not fit into any then-existing alternative theories of causation, the *Sindell* court noted that changing technology had led to the creation of products that could harm consumers but could not be traced to any specific producer, and opted to adapt the rules of causation and liability to better account for such products.[145] Thus, the court required defendants to prove that they were not responsible for producing the drug ingested by the plaintiff's mother.[146] Only one of the 11 defendants was able to do so. The court apportioned damages among the 10 remaining defendants based on their market share of DES during the relevant time frame.

Several jurisdictions in addition to California have adopted the market share liability theory in the DES or some other drug-related context.[147] For the most part, however, even courts in jurisdictions embracing the theory have declined to extend market

---

139. Tobacco Master Settlement Agreement, Cal. Dep't of Justice, *available at* http://oag.ca.gov/tobacco/msa.
140. 26 Cal. 3d 588 (1980).
141. *Id.* at 593–94.
142. *Id.* at 594–95.
143. Sindell v. Abbott Labs., 149 Cal. Rptr. 138 (Ct. App. 2d 1978).
144. *Sindell*, 26 Cal. 3d 588.
145. *Id.* at 610–13.
146. *Id.* at 612–13.
147. *See, e.g., id.*; Conley v. Boyle Drug Co., 570 So. 2d 275 (Fla. 1990) (DES products liability claim); Smith v. Cutter Biological, Inc., 823 P.2d 717 (Haw. 1991) (claim against manufacturer of blood shield product from which plaintiff contracted AIDS); Hymowitz v. Eli Lilly & Co., 539 N.E.2d 1069 (N.Y. 1989) (DES products liability claim); Martin v. Abbott Labs., 689 P.2d 368 (Wash. 1984) (DES products liability claim); Morris v. Parke, Davis & Co., 667 F. Supp. 1332, 1342 (C.D. Cal. 1987) (DPT vaccine products liability claim); McElhaney

share liability to other types of cases. For example, multiple courts have refused to apply market share liability to lead or asbestos cases,[148] focusing on the non-fungibility (i.e., the lack of functional interchangeability and/or physical distinguishability) of those substances.[149] In DES cases, courts noted that because DES was fungible, "there was no difference between the risks associated with the drug as marketed" by various companies; all DES presented the same risk of harm.[150] Thus, "there was no inherent unfairness in holding the companies accountable based on their share of the DES market."[151] In contrast, there are multiple varieties of asbestos fibers with differing harmful effects that are incorporated into asbestos products in varying quantities by manufacturers.[152] Lead has been deemed non-fungible for similar reasons.[153]

In the context of climate change, if a plaintiff can demonstrate that GHG emissions are fungible,[154] that an injury was caused by GHG emissions, and that it has joined in the action a "substantial percentage" of the companies responsible for those emissions, the causation burden could shift, forcing a defendant seeking to avoid liability to prove that it was not responsible for the emissions leading to the alleged harm.[155] If a court were to apply these principles in climate change tort litigation, defendants could be liable for damages proportional to their market share responsibility for emissions. Plaintiffs likely will advocate for the application of the market share liability theory in coming years.

---

v. Eli Lilly & Co., 564 F. Supp. 265 (D.S.D. 1983) (applying South Dakota law); Collins v. Eli Lilly & Co., 342 N.W.2d 37 (Wis. 1984), *cert. denied*, 469 U.S. 826 (1984) (DES products liability claim).

148. *See, e.g.,* In re Related Asbestos Cases, 543 F. Supp. 1152 (N.D. Cal. 1982); Celotex Corp. v. Copeland, 471 So. 2d 533, 537–39 (Fla. 1985); Sholtis v. Am. Cyanimid, 568 A.2d 1196, 1203–05 (N.J. Super. Ct. App. Div. 1989).

149. In declining to extend the market share liability theory, courts also have noted the difficulty in defining the market for various substances, as well as the lack of a "signature injury" that can be traced to a particular substance. *See, e.g.,* Goldman v. Johns-Manville, 514 N.E.2d 691, 700, 701 (Ohio 1987) (finding the application of market share liability theory inappropriate in asbestos cases because (1) asbestos is not a single fungible product, but rather a general name for a family of products, (2) defining the relevant product and geographic markets would be difficult because asbestos is put to many uses, and (3) the extended time frame associated with potential exposures); Santiago v. Sherwin-Williams Co., 782 F. Supp. 186, 192–95 (D. Mass. 1992) (finding the application of market share liability theory inappropriate in lead cases because (1) ingestion of lead does not lead to "signature injury" that can be traced back to lead, (2) defining the relevant market would be difficult due to breadth of lead pigment distribution, and (3) defendants were bulk suppliers rather than manufacturers of lead).

150. *Goldman*, 514 N.E.2d at 701.

151. *Id.*

152. *See* In re Related Asbestos Cases, 543 F. Supp. at 1158.

153. *See* Skipworth v. Lead Indus. Ass'n, Inc., 690 A.2d 169, 173 (Pa. 1997) (finding lead pigments not fungible because they "had different chemical formulations, contained different amounts of lead, and differed in potential toxicity").

154. Although GHG emissions do not fit neatly into the fungibility framework established by the DES cases, some courts may prove willing to use CO_2e to establish a fungible metric for market share liability purposes. EPA is using a CO_2e analysis to distinguish among the varying contributions of different GHGs.

155. *See Sindell*, 26 Cal. 3d at 612 (by analogy).

c. Foreseeability

Although not strictly a climate change-related tort case, the decision in *In re Katrina Canal Breaches Consolidated Litigation*[156] suggests a potential construct for holding private entities in charge of certain types of structures liable for injuries caused by "the decay or obsolescence of [such structures] due to erosion, sea level rise, and other ongoing conditions, whether of natural or human origin."[157] In this particular case, hundreds of New Orleans–area property owners sued to recover for flooding damages incurred in the aftermath of 2005's Hurricane Katrina, many naming the federal government as a defendant. The allegations against the government centered on the alleged negligence of the Army Corps of Engineers in dredging the Mississippi River Gulf Outlet (MRGO), a shipping channel between New Orleans and the Gulf of Mexico, as well as various levees located next to the channel and around the city.[158] In particular, the plaintiffs pointed to the Corps' alleged negligence in maintaining the channel in the face of certain known or knowable hydrological risks, purportedly causing levees to fail and aggravating Katrina's effects on the city and surrounding areas.[159]

Following a 19-day trial involving seven of the plaintiffs, the presiding judge issued a 156-page decision finding "gross negligence" by the corps in its operation of MRGO and awarding just under $720,000 to five of the plaintiffs.[160] Among other things, the decision left open the possibility of trials by hundreds of the remaining plaintiffs and the potential for hundreds of millions of dollars of additional liability to the corps.[161] In its March 2, 2012, decision upholding the district court's ruling, the U.S. Court of Appeals for the Fifth Circuit found persuasive—among other things—the fact that substantial scientific information available to the corps revealed at least some of the dangers created by their actions and omissions with regard to MRGO's maintenance.[162]

As Michael B. Gerrard, Director of the Center for Climate Change Law at Columbia University, has since pointed out, the Fifth Circuit's decision in *In Re Katrina Canal Breaches Consolidated Litigation* could potentially affect the liability of private entities responsible for maintaining structures, like MRGO, that fail to withstand extreme weather events such as a hurricane: "The MRGO litigation is an example of how a property manager was found liable for ignoring scientific evidence of perils it faced; the fact that the property manager was a federal agency does not diminish the

---

156. 673 F.3d 381 (5th Cir. 2012).
157. Michael B. Gerrard, *Hurricane Katrina Decision Highlights Liability for Decaying Infrastructure*, 247 N.Y. L.J., May 10, 2012.
158. 673 F.3d at 385.
159. *Id.*
160. *In re* Katrina Canal Breaches Consol. Litig., 647 F. Supp. 2d 644 (E.D. La. 2009).
161. *See* Gerrard, *supra* note 157.
162. 673 F.3d at 394–96. This portion of the court's decision rejected application of the discretionary function exception to the Federal Tort Claims Act to the plaintiffs' claims against the federal government.

case's relevance to a negligence analysis involving private parties."[163] Under the rationale articulated by the court, the proliferation of scientific information available about emerging risks associated with climate change could potentially lead to a conclusion that certain types of conditions were "foreseeable" for purposes of any negligence analysis.[164] Some of the scientific studies and sources of information relating to the effects of climate change are discussed in Chapters 2 and 4.

B. Climate Change-Related Insurance Coverage Litigation: An Introduction to *AES Corp. v. Steadfast Insurance Co.*

In 2008, Steadfast Insurance Company filed what may be the first declaratory judgment action regarding insurance coverage for climate change-related claims.[165] Commentators referred to *AES Corp. v. Steadfast Insurance Co.* as a major test case and an indicator that "the battle has begun" on insurance coverage for climate suits.

In *AES*, Steadfast asserted that it had no duty to defend its insured, AES, in the underlying *Kivalina v. ExxonMobil Corp.* action for three primary reasons.[166] First, Steadfast argued that the release of GHGs does not constitute a covered "occurrence" because the release and resulting climate change effects are not accidental but rather foreseeable impacts of the business of producing energy. Second, Steadfast argued that the release of GHGs falls under the pollution exclusion in part because the U.S. Supreme Court, in *Massachusetts v. EPA*, recognized certain GHGs as "pollutants" under the Clean Air Act. Finally, Steadfast argued that the release of GHGs falls under the policies' loss in progress exclusion because such emissions began before the effective date of the at-issue policies.

In an initial victory for insurers, the Virginia Supreme Court held that insurance companies do not have to defend utility companies accused of intentional wrongdoing in connection with climate change liability lawsuits.[167] The court concluded that the underlying climate-change claims in the *Kivalina* lawsuit did not constitute an "occurrence" under AES's CGL policies.[168] Because the court decided the case on the occurrence issue, the court did not reach the issue of whether the pollution exclusion might apply. After agreeing to rehear the case, the Virginia Supreme Court reached the same

---

163. Gerrard, *supra* note 157.
164. See id.
165. Complaint for Declaration Relief, Steadfast Ins. Co. v. AES Corp., No. 2008-858 (Va. Cir. Ct. Arlington County July 9, 2008).
166. These bases are discussed in more detail in Chapter 8, Sections B.1, C.1.b(2), and D.3.
167. AES Corp. v. Steadfast Ins. Co., 715 S.E.2d 28 (Va. 2011) *petition for reh'g granted* (Jan. 17, 2012).
168. Id.

conclusion—that there was no "occurrence" and thus Steadfast had no duty to defend under the CGL policies.[169]

The *AES* case could have important implications for both insurers and companies with potential exposure to climate change-related tort claims. Although policyholders and their counsel are likely to now press coverage issues in more favorable jurisdictions, the initial decision nonetheless was one step toward resolving the question of whether such claims are covered under CGL policies. Future coverage litigation in this area in other states could also focus on the occurrence issue, as well as on the pollution exclusion and known loss issues not reached in the *AES* case. Even when coverage cases like *AES* are successful, the transactional costs associated with these climate change-related coverage disputes could be significant. See Chapter 8, Sections B.1, C, and D.3 for detailed discussions of the occurrence, known loss, and pollution exclusion as liability drivers and the arguments set forth in the *AES* litigation.

C. Public Trust Litigation

On May 4, 2011, children's, young-adult, and various environmental groups began suing the federal government and the 50 states for violations of the public trust doctrine in various actions across the country.[170] Our Children's Trust, an Oregon nonprofit, is coordinating the litigation.[171] According to the plaintiffs, "the public trust doctrine is an ancient legal mandate establishing a sovereign obligation in states to hold critical natural resources in trust for the benefit of their citizens."[172] They claim that the federal government and the states have not properly protected the atmosphere—a resource that they hold in trust for present and future generations—from GHG emissions that lead to climate change. Plaintiffs request in the New Mexico complaint, for example, that the court enter a judgment that "(1) the public trust doctrine is operative in New Mexico and, pursuant to this doctrine, the State holds the atmosphere in trust for the public; (2) the State has an affirmative fiduciary duty to establish and enforce limitations on the levels of greenhouse gas emissions as necessary to protect and preserve the public trust in the atmosphere; (3) the State's fiduciary duty to protect the atmospheric trust is defined by the best available science; and (4) the State has breached its fiduciary duty

---

169. AES Corp. v. Steadfast Insurance Co., No. 100764, slip op. (Va. Apr. 20, 2012).

170. *See, e.g.,* Complaint, Alec L. v. Jackson, No. 11-2203 (N.D. Cal. May 4, 2011) (public trust action against the federal government); Petition for Original Jurisdiction, Barhaugh v. State, No. OP 11-0258 (Mont. May 4, 2011).

171. *Legal Action*, Our Children's Trust, http://ourchildrenstrust.org/page/31/legal-action (providing information on the status of the lawsuits).

172. *See, e.g.,* Compl. for Declaratory Relief ¶ 23, *Sanders-Reed, et al. v. Martinez*, No. D-0101-CV-2011-10514 (N.M. 1st Dist. Ct. May 4, 2011).

to protect the public trust in the atmosphere by failing to exercise its right of control over the atmosphere in a manner that promotes the public's interest in the atmosphere and does not substantially impair this resource."[173] Many of the lawsuits also argue that the public trust doctrine is inherent in particular provisions in state constitutions.[174] In the complaint against the federal government, plaintiffs request that the government act immediately to reduce GHG emissions by 6 percent a year.[175]

The state public-trust litigation largely has favored defendants thus far, but plaintiffs have survived motions to dismiss in some state cases. For instance, in June 2011, the Montana Supreme Court denied a petition seeking enforcement of a state constitutional obligation to regulate GHGs in the atmosphere because it concluded the court lacked original jurisdiction.[176] In January 2012, a Minnesota district court dismissed a lawsuit seeking a declaration that the atmosphere is protected by the public trust doctrine.[177] The court held that the public trust doctrine only applies to navigable waters, not the atmosphere.[178] In Alaska, the superior court dismissed a public trust action similar on grounds that the public trust doctrine does not include the atmosphere and, moreover, that plaintiffs' claims constitute nonjusticiable political questions.[179]

The New Mexico case, however, has survived dismissal and remains active as of the time of this writing. On June 29, 2012, in a ruling from the bench, the presiding judge denied the New Mexico state defendants' motion to dismiss, thereby allowing the case to go forward.[180] Thereafter, on July 9, 2012, a Texas judge rejected arguments by the Texas Commission on Environmental Quality that only water is a "public trust," finding that the atmosphere and air also must be protected for public use.[181] It remains to be seen whether the Texas judge's ruling will be adopted in any of the

---

173. Compl. ¶ 8, *supra* note 172.

174. *See, e.g., id.* ¶ 27; *see also* Petition for Original Jurisdiction, Barhaugh v. State, No. OP 11-0258 (Mont. May 4, 2011). The petition cites the Montana Constitution article IX, section 1 ("The state . . . shall maintain . . . a clean and healthful environment in Montana for present and future generations.") and article II, section 3 ("All persons are born free and have certain inalienable rights. They include the right to a clean and healthful environment. . . .").

175. Complaint, Alec L. v. Jackson, No. 11-2203 (N.D. Cal. May 4, 2011).

176. Order, Barhaugh v. State, No. OP 11-0258 (Mont. June 15, 2011).

177. Memorandum and Order, Aranow v. Minnesota, File No. 62-CV-11-3952 (Minn. Dist. Ct. Jan. 30, 2012).

178. *Id.*

179. Order Re: Motion to Dismiss, Kanuk v. Alaska, Case No. 3AN-11-07474CI (Alaska Super. Ct. Mar. 16, 2012).

180. April Reese, *N.M. Judge Allows "Public Trust Resource" Case to Proceed*, available at http://www.wildearthguardians.org/site/DocServer/NM_judge_allows_public_trust_resource_case_to_proceed.pdf?docID=5622&AddInterest=1058 (June 29, 2012).

181. Ramit Plushnick-Masti, *Texas judge rules atmosphere, air is public trust*, DALLAS MORNING NEWS, July 11, 2012, *available at* http://www.dallasnews.com/news/state/headlines/20120711-texas-judge-rules-atmosphere-air-is-public-trust.ece.

state cases that remain pending (or in any appeals of cases that have been dismissed), but it is clear that some plaintiffs intend to rely on the ruling as persuasive authority.[182]

After an initial venue change, the federal public trust litigation has been dismissed. In December 2011, the U.S. District Court for the Northern District of California transferred the lawsuit to the U.S. District Court for the District of Columbia on grounds that the lawsuit challenged broad, nationwide policies that are prepared by federal agencies in the nation's capital.[183] On June 1, 2012, the U.S. District Court for the District of Columbia dismissed the action on grounds that the public trust doctrine is a matter of state law, not federal law.[184] In addition, the court held that even if the public trust doctrine were a federal common law claim, such a claim has been displaced by the Clean Air Act.[185]

The adjudication of the public trust cases will be another interesting chapter in the climate change–common law litigation debate. Given that these cases are filed against governmental entities and so far have had limited success, they generally do not present a direct risk to insurers because their insureds are not involved in these claims. The new public trust litigation, however, is illustrative of the potential for new theories to emerge just as others are being rejected. These cases were filed around the time that the Supreme Court rejected the federal common law of nuisance as a vehicle to obtain redress for alleged climate change-related injuries. New theories and targets may continue to emerge as plaintiffs weigh the costs associated with climate change.

D. Claims for Natural Resource Damages Pursuant to CERCLA

Another area of potential litigation on the climate change horizon is claims for NRD under CERCLA. Although actions seeking recovery for environmental damages associated with climate change are unlikely to succeed under existing CERCLA regulations, increased regulatory activity in this area may make NRD claims more feasible. Regardless of their ultimate potential for success, these claims would probably result in increased coverage litigation as insureds seek refuge and defense under applicable insurance policies.

Congress enacted CERCLA in 1980 in order to provide for the cleanup of hazardous substances emitted into the environment.[186] Although the act has several components, a primary aspect of the legislation permits "trustees" to recover damages for injuries to

---

182. *See id.*
183. Order Granting Defendants' Motion to Transfer Venue, Alec L. v. Jackson, No. 3:11-cv-02203 EMC (N.D. Cal. Dec. 6, 2011).
184. Alec L. v. Jackson, 1:11-cv-02235 RLW, 2012 WL 1951969, at *3–4 (D.D.C. May 31, 2012).
185. *Id.* at *4–5.
186. 42 U.S.C. §§ 9601 *et seq.*

"natural resources" caused by the release of a "hazardous" substance.[187] Natural resources are broadly defined to include "land, fish, wildlife, biota, air, water, ground water, drinking water supplies, and other such resources."[188] Only resources belonging to, managed by, held in trust, or otherwise controlled by the United States, any state or local government, a foreign government, or any American Indian tribe fall within the purview of the statute, and only the trustees of such lands are empowered to bring NRD actions.[189]

No NRD actions seeking recovery of climate change-related damages have been filed to date, but recent climate change litigation grounded in more traditional environmental tort theories is starting to echo NRD actions. In particular, the nature of the allegations of harm in the *Kivalina* lawsuit suggests the potential for NRD claims to arise in the near future. The *Kivalina* plaintiffs, members of a native Alaskan tribe, sued numerous power, utility, and oil companies and alleged that defendants' GHG emissions caused global warming, which in turn caused erosion of the Kivalina coastline due to thinning of sea ice.[190] The *Kivalina* plaintiffs did not seek NRD under CERCLA, but the plaintiffs could potentially qualify as trustees of natural resources.

It is likely that trustees are hesitating to bring NRD actions targeted at the effects of climate change because of the difficulty of satisfying two major elements of an NRD claim. First, there must have been a release of a "hazardous" substance. Second, plaintiffs must prove that the release is causally related to the alleged NRD. Under current regulations and available science, these requirements are major roadblocks to a successful NRD claim. These hurdles, however, may soon become less difficult to surmount.

At this time, no GHGs are currently listed as "hazardous" substances pursuant to CERCLA or any other federal statutes as is required by section 101(14) of CERCLA. EPA, however, found that the mixture of six key GHGs in the atmosphere may be expected to endanger public health and welfare ("endangerment finding").[191] This finding may bring trustees one step closer to an NRD claim that is likely to withstand motions to dismiss at the pleadings stage of litigation.

The causal impediment to climate-based NRD claims also may be eroding. GHGs, in particular CO_2, are ubiquitous in nature, and attempting to distinguish the harm caused by one set of emissions over another is problematic. As the science continues to

---

187. *Id.* § 9607(a)(4)(C).
188. *Id.* § 9601(16).
189. *Id.* Citizen suits for NRD are not permitted under CERCLA. *See, e.g.*, Alaska Sport Fishing Ass'n v. Exxon Corp., 34 F.3d 769, 772 (9th Cir. 1994) (holding that trustees, not private parties, have the authority to recover for NRD including lost-use damages).
190. Complaint for Damages Demand for Jury Trial ¶¶ 3–4, Native Vill. of Kivalina v. ExxonMobil Corp., No. 4:08-cv-01138-SBA (N.D. Cal. Feb. 26, 2008). *See* Chapter 6.A.3 (discussing the *Kivalina* case in detail).
191. *See* Endangerment and Cause or Contribute Findings for Greenhouse Gases Under Section 202(a) of the Clean Air Act, 74 Fed. Reg. 66,496 (Dec. 15, 2009).

evolve and as more entities, including federal and state legislators, public interest groups, and others, begin to funnel more money into these endeavors, it may become easier for plaintiffs to develop a causal connection between the alleged hazardous emission and particular types of harm. For example, EPA is using a CO_2 equivalent (CO_2e) analysis for comparing the roles of various GHGs. Alternatively, courts may not await scientific evidence or regulatory constructs and may begin eroding the traditional legal concept of causation. There is precedent in the NRD context that the causal nexus between an entity's hazardous release and the specific hazardous materials that cause injury does not have to be perfect.[192] Thus, plaintiffs might try to use CERCLA NRD as a vehicle instead of tort claims if causation issues in tort cases prove to be problematic.

With respect to underlying CERCLA NRD claims, insureds will look to environmental liability and CGL policies for coverage. See Chapter 7.A.3.b–c. Regardless of the ultimate ability of trustees to prevail in NRD actions based on climate-change effects, the mere filing of such lawsuits may trigger an insurer's duty to defend and cause insurers to incur defense costs.

E. Claims Arising out of Corporate Disclosure and Management of Climate Change Risk

> Increasing demands for voluntary climate change disclosure, along with the development and enforcement of required disclosures, could have implications for insurers' enterprise and potential claims liability exposure.

In recent years, investors have demanded greater disclosure of climate change-related risk. The evolving voluntary and regulatory mechanisms for climate change disclosure could have potential implications for insurers' enterprise and potential claims liability exposure with respect to climate change. As for direct liability, insurers could face exposure due to their own climate change-related statements. As to potential claims liability, insurers could begin to receive claims because of the climate change disclosure practices of its insureds. For example, disclosure irregularities could subject insureds to regulatory or shareholder claims related to concealment, misrepresentation, and mismanagement of climate change-related risk, which has implications for D&O coverage.[193] The data from disclosures also may serve as fodder for a host of

---

192. *In re* Acushnet River & New Bedford Harbor, 722 F. Supp. 893, 897 (D. Mass. 1989); Coeur D'Alene Tribe v. Asarco, Inc., 280 F. Supp. 2d 1094, 1124 (D. Idaho 2003).

193. The implications for D&O coverage are discussed in Chapter 7.A.3.a.

other climate change-related claims, which has implications for CGL and environmental liability exposure.[194]

This section provides extensive background on trends in shareholder resolutions related to climate change, potentially applicable SEC disclosure requirements, the 2010 SEC Guidance Regarding Climate Change and its effect on disclosure requirements, New York regulatory actions related to climate change risk disclosure, and many of the existing voluntary climate change disclosure programs. (The NAIC Climate Risk Disclosure Survey, which is applicable to insurers only in certain states, is discussed in Chapter 5.H.1.a–c.)

1. Shareholder Resolutions Related to Climate Change

Shareholders have not yet filed litigation aimed at corporate climate change-related activities. Shareholders are, however, demanding corporate action on climate change and access to information necessary to support potential securities claims related to climate change. Shareholders have prepared resolutions related to climate change and GHG emissions at a higher rate in recent years.[195] Ceres, self-described as "a national coalition of investors, environmental organizations and other public interest groups working with companies and investors to address sustainability challenges such as global climate change,"[196] reported that a then record 54 shareholder proposals on climate change were submitted to U.S. companies for the 2008 proxy season—a number nearly double the amount filed in 2006.[197] In the years that followed, according to Ceres, the number of climate change-related shareholder resolutions continued to increase steadily, with 68 such resolutions filed during the 2009 proxy season, 96 filed during the 2010 proxy season, and 111 filed in 2011.[198]

---

194. The implications for CGL and environmental liability coverage are discussed in Chapter 7.A.3.b–c.
195. A shareholder resolution is a formal request to a company's management on a particular issue that is subject to shareholder vote at the annual shareholder's meeting.
196. Shortly after the 1989 Exxon-Valdez oil spill in Alaska's Prince William Sound, a group of investors launched Ceres, a nonprofit organization aimed at tackling the environmental and social impacts of business operations, including global climate change. Over the past 20 years, the organization's membership has grown to include a national network of investors, environmental organizations, and other public interest groups managing approximately seven trillion dollars in assets whose mission is "integrating sustainability into capital markets for the health of the planet and its people." The California Public Employees Retirement System (CalPERS), which is the nation's largest public pension fund, is a leading member of Ceres. *See* Ceres, http://www.ceres.org (last visited Jan. 31, 2012).
197. Ceres, Investors File Record Number of Global Warming Resolutions with U.S. Companies (Mar. 6, 2008).
198. *Corporate Climate Change-related Risk on Shareholders' Minds*, Ceres (Mar. 14, 2010), http://www.ceres.org/press/press-clips/corporate-climate-change-related-risk-on-shareholders2019-minds; *2011 Proxy Season Report: Fracking, Water Scarcity, Other Issues Show Shareholders Resolute on Climate & Related Sustainability Resolutions*, Ceres (June 23, 2011), http://www.ceres.org/press/press-releases/2011-proxy-season-report-fracking-water-scarcity-other-issues-show-shareholders-resolute-on-climate-sustainability-resolutions;

As shareholder resolutions related to climate change increase, shareholder support of climate change resolutions also appears to be on the rise.[199] In 2007, climate change resolutions received, on average, 18.7 percent shareholder support (up from 10.8 percent in 2005) while shareholder resolutions requesting GHG emissions reports averaged 30.6 percent shareholder support.[200] Among the investors tracked by Ceres from 2002 to 2012, average voting support for similar resolutions rose during the course of the decade to a rate of 24 percent, with a third of the resolutions going to a vote garnering at least 30–40 percent support and a few achieving majority support.[201] These developments may be considered in underwriting D&O coverage. See Chapter 7.A.3.a, for further discussion of the implications of climate change issues for D&O.

Generally, climate-change shareholder resolutions seek voluntary action by a corporation's management to (1) disclose more information about its GHG emissions, (2) set goals and timetables to achieve absolute reductions in GHG emissions from operations and products, and (3) analyze risks and opportunities created by climate change.[202] Notwithstanding the foregoing, the specific focus of climate change resolutions varies by industry. Exemplar proposals considered by companies in the automotive, oil and gas, energy, retail, and financial services industries during the 2008–2011 proxy seasons range from requests to adopt GHG emissions reduction goals (filed against, e.g., an automotive company, petroleum refining companies, food-processing companies, and a major insurance company) to requests for reports on global warming and/or climate change impacts and strategies (filed against, e.g., a retail grocery chain, an electronics and other electrical equipment manufacturer, and a securities brokerage firm).

Shareholder Successes on Climate, Energy & Sustainability, CERES (Feb. 2012), http://www.ceres.org/files/in-briefs-and-one-pagers/proxy-power-shareholder-successes-on-climate-energy-sustainability/view. Ceres maintains a searchable database of shareholder resolutions filed by its investors on various sustainability-related issues, including climate change, energy, and sustainability reporting. *See Shareholder Resolutions*, CERES, http://www.ceres.org/incr/engagement/corporate-dialogues/shareholder-resolutions (last accessed Mar. 27, 2012).

199. BETH YOUNG, CELINE SUAREZ & KIMBERLY GLADMAN, CERES, CLIMATE RISK DISCLOSURE IN SEC FILINGS 7 (June 2009).

200. *Id.*

201. *Shareholder Successes on Climate, Energy & Sustainability*, CERES (February 2012), http://www.ceres.org/files/in-briefs-and-one-pagers/proxy-power-shareholder-sucesses-on-climate-enerty-sustainability/view.

202. GAURAV SHIL & KATHERINE N. BLUE, GREENHOUSE GAS EMISSIONS: A CASE STUDY OF DEVELOPMENT OF DATA COLLECTION TOOL AND CALCULATION OF EMISSIONS 2–3, *available at* http://www.epa.gov/ttnchie1/conference/ei16/session3/shil.pdf; ERNST & YOUNG LLP, SHAREHOLDERS PRESS BOARDS ON SOCIAL AND ENVIRONMENTAL RISKS: IS YOUR COMPANY PREPARED? 2 (2011), *available at* http://www.ey.com/Publication/vwLUAssets/CCaSS_social_environmental_risks/$FILE/CCaSS_social_environmental_risks.pdf.

During the 2008 proxy season, several companies succeeded in omitting climate change-related shareholder proposals from company proxy statements following challenges to the SEC under SEC Rule 14a-8(i).[203] For example:

- Ford Motor Company successfully petitioned the SEC to omit proposals to improve fuel economy and to report on and reduce GHG emissions.[204]
- Motors Liquidation Company (formerly General Motors) successfully petitioned the SEC to omit proposals to improve fuel economy and to report on and reduce GHG emissions.[205]
- Bank of America successfully petitioned the SEC to omit a proposal to cease financing coal operations.[206]

While some companies have enjoyed success under Rule 14a-8(i), others have not fared as well. General Electric Co. (GE), for example, received a shareholder proposal in 2008 requesting that GE prepare a global warming report.[207] The proposed global warming report would require GE to describe (1) the specific scientific data and studies relied on to formulate GE's climate policy; (2) the extent to which GE believes human activity will significantly alter the global climate, whether such change is necessarily undesirable and whether a cost-effective strategy for mitigating any undesirable change is practical; and (3) the estimates of costs and benefits to GE of its climate policy.[208] GE appealed to the SEC under Rule 14a-8(i)(7) alleging that the resolution dealt with matters relating to GE's "ordinary business" operations as it related to GE's evaluation of the risks and benefits of aspects of GE's business operations, and thus GE could exclude the proposal from the proxy statement.[209] In its response, the SEC did not agree with GE's view and, consequently, it did not believe that GE was permitted to exclude the proposal under Rule 14a-8(i)(7).[210]

---

203. Rule 14a-8(i) permits companies to exclude shareholder proposals from their proxy materials if the shareholder proposal (1) fails to comply with the Rule's procedural requirements or (2) falls within one of thirteen substantive bases for exclusion provided by the Rule. 17 C.F.R. § 240.14a-8(i).
204. Ford Motor Co., SEC No-Action Letter (Feb. 29, 2008), *available at* http://www.sec.gov/Archives/edgar/vprr/08/9999999997-08-008920.
205. Motors Liquidation Co. (formerly General Motors Corp), SEC No-Action Letter (Mar. 13, 2008), *available at* http://www.sec.gov/Archives/edgar/vprr/08/9999999997-08-011192.
206. Bank of America Corp., SEC No-Action Letter (Feb. 25, 2008), *available at* http://www.sec.gov/Archives/edgar/vprr/08/9999999997-08-008642.
207. General Electric Co., SEC No-Action Letter (Jan. 15, 2008), *available at* http://www.sec.gov/Archives/edgar/vprr/08/9999999997-08-003084.
208. *Id.*
209. *Id.*
210. *Id.*

On October 27, 2009, the SEC issued revised guidance on Rule 14a-8(i)(7).[211] In applying this rule, SEC had allowed companies to exclude shareholder proposals that related to assessment of risk. In light of the number of no-action requests on resolutions relating to environmental, health, and financial risk in recent years, however, SEC decided to revise its guidance. SEC was concerned that application of its existing analytical framework may have resulted in unwarranted exclusion of proposals that relate to the evaluation of risk but nonetheless relate to significant corporate policy issues. Thus, on a prospective basis, the fact that a proposal would require a risk assessment will not be a basis for exclusion. Although the guidance does not mention climate change specifically, it could have significant ramifications in the climate change context. Ceres applauded the action and argued that this change will allow investors to expressly inquire about the impacts of climate change on a company.[212]

Generally, management opposes shareholder proposals that are considered for shareholder vote, and the company will recommend that the shareholders vote against proposals in the proxy statement. The effectiveness of this resistance was evident in the 2008 proxy season. Despite the fact that the number of climate change resolutions submitted in 2008 for shareholder vote was record setting, none of these resolutions received majority shareholder support. The 2009 proxy season, however, saw the first majority shareholder vote in favor of a climate change resolution when the shareholders of IDACORP, Inc. voted by 52 percent for the establishment of GHG reduction goals.[213]

Although shareholder resolutions generally are not binding, a company's failure to implement a proposal that receives a large percentage of shareholder support can harm the company's reputation and erode shareholder support for directors. Indeed, 39.5 percent of Exxon's investors voted in favor of a resolution proposed by the Rockefeller family to split the role of the chief executive officer and the chairman of the board of directors, in part to stimulate greater debate on the company's response to climate change.[214] Similarly, a resolution to limit the company's GHG emissions secured 30.98 percent support, while a proposal to increase investment in renewable energy won 27.4 percent. Exxon's shareholder support has remained relatively steady to date.

---

211. SEC Staff Legal Bulletin No. 14E (CF), Shareholder Proposals (Oct. 27, 2009) *available at* http://www.sec.gov/interps/legal/cfslb14e.htm.

212. Press Release, Ceres, Ceres Applauds SEC Decision Allowing Financial Risks in Environmental and Social Resolutions (Oct. 28, 2009), *available at* http://www.ceres.org/press/press-releases/ceres-applauds-sec-decision-allowing-financial-risks-in-environmental-and-social-resolutions.

213. IDACORP, Inc., Form 10-Q for the Quarterly Period ended June 30, 2009, *available at* http://www.sec.gov/Archives/edgar/data/49648/000105787709000090/esa10q.htm.

214. Exxon Mobil Corp., Form 10-Q for the Quarterly Period ended June 30, 2008, *available at* http://www.sec.gov/Archives/edgar/data/34088/000003408808000104/r10q080508.htm.

During the 2011 proxy season, 26.5 percent of shareholders supported the proposal on setting GHG emission goals.[215] These results confirm that a sizable minority of Exxon's shareholders oppose the company's traditionally conservative approach to climate change and alternative energy.

Shareholder interest in climate change is likely to continue to increase as climate change continues to be a high-profile issue and EPA, Congress, and certain states regulate or move toward regulating GHG emissions.

2. Potential Litigation Related to Misrepresentation, Concealment, or Mismanagement of Climate Change Risk

In the future, directors and officers may face both regulatory and shareholder actions if they misrepresent or conceal climate change-related risks or mismanage climate change-related issues. Directors and officers in energy-intensive sectors will be particularly vulnerable to such claims given that these sectors have high GHG emissions. These types of claims could have potentially significant implications for D&O coverage.

In order to fully understand potential climate change-related disclosure liability, the current and evolving mandatory disclosure requirements are discussed below. Voluntary climate-change disclosure programs also are discussed below because disclosures through these programs may form the basis for future climate change mismanagement claims.

3. Role of the SEC in Climate Risk Disclosure

a. Background: Petitions to the SEC for Further Regulation of Climate Change Risk Disclosure

Some investors and other interested parties have expressed concern that businesses were underreporting climate change risks to the SEC. A 2009 report issued by investor NGO Ceres in cooperation with other environmental organizations, for example, described "an alarming pattern of non-disclosure by corporations regarding climate risks."[216] According to that study, 76.3 percent of the SEC annual reports filed by S&P 500 corporations in 2008 failed to mention climate change risks or opportunities, and less than 10 percent of companies in the financial services sector discussed climate change in their 2008 Form 10-K.[217]

---

215. PROXY MONITOR, http://proxymonitor.org/ (last visited June 7, 2011).
216. *See, e.g.*, KEVIN L. DORAN, ELIAS J. QUINN & MARTHA G. ROBERTS, CERES, RECLAIMING TRANSPARENCY IN A CHANGING CLIMATE: TRENDS IN CLIMATE RISK DISCLOSURE BY THE S&P 500 FROM 1995 TO THE PRESENT (2009), *available at* http://www.ceres.org/resources/reports/reclaiming-transparency-in-a-changing-climate-1/view (this download requires registration).
217. *Id.*

Investors have been demanding that companies include more climate change risk disclosures in their periodic SEC reports. The SEC received multiple requests from investor groups either to strengthen current disclosure requirements by construing climate change as a material environmental risk or to issue separate disclosure mandates that specifically target climate change-related risks.

(1) Ceres September 2007 Petition

In September 2007, Ceres filed a groundbreaking petition asking the SEC to require publicly traded companies to fully disclose their financial risks pertaining to climate change.[218] Ceres pointed out variations in the quality of disclosure among SEC filings of members of the auto, insurance, energy, petrochemical, and utilities industries from 2001 to 2006, calling it an "inconsistent patchwork of disclosure."[219] The petition emphasized that current disclosure practices leave investors "in the dark" about the financial implications of environmental issues such as climate change.[220]

The core concept underlying the petition was that climate change is a "known trend" or, at a minimum, a known uncertainty that will have a material effect on business operations and investor decision making.[221] The petition asked the SEC to confirm, after a "close and well informed review" of relevant information, that registrants are required to disclose material climate change-related risks under existing disclosure policies.[222] Ceres hoped that clarification from the SEC on the reach of current regulations would both enlarge the number of public companies addressing climate change in their SEC reporting as well as enhance the quality of the companies' climate change disclosure practices.

(2) June 12, 2009 Petition

On June 12, 2009, a coalition of investors supplemented the Ceres petition by appealing, once again, to the SEC to issue an interpretive release on climate change disclosures.[223] The letter's 41 signatories include some of the United States' largest public pension funds, state treasurers, asset managers, foundations, and other institutional investors with approximately $1.4 trillion in assets under their management. The June petition asked the SEC to clarify that material climate-related information must

---

218. *See* Petition for Interpretive Guidance on Climate Risk Disclosure, at 2 (2007), *available at* http://www.sec.gov/rules/petitions/2007/petn4-547.pdf [hereinafter *Climate Risk Petition*].
219. *Id.* at 47.
220. *Id.* at 48.
221. *Id.* at 8.
222. *Id.* at 53.
223. Press Release, Investor Network on Climate Risk, *Investors with $1.4 trillion in Assets Call on the SEC to Improve Disclosure of Climate Change and Other Risks* (June 12, 2009), *available at* http://www.ceres.org/incr/news/investors-with-1.4-trillion-in-assets-call-on-the-sec-to-improve-disclosure-of-climate-change-and-other-risks.

be included in SEC filings under existing law and to review the adequacy of climate risk–related disclosures. In particular, the June 12 letter urged the SEC to:

- strengthen current disclosure requirements by issuing formal interpretative guidance on the materiality of risks posed by climate change;
- enforce existing disclosure requirements for material environmental risks;
- permit shareholders to submit resolutions at corporate meetings related to climate change; and
- require disclosure of material environmental risks based on the Global Reporting Initiative, an existing standard used by more than 1,000 companies to measure and report their economic, environmental, and social performances, including measurements of GHG emissions, labor standards and human rights.

In response to these and other requests, the SEC issued Commission Guidance Regarding Disclosure Related to Climate Change in 2010 (2010 SEC Guidance) addressing some but not all of these requests.[224] Current SEC laws and regulations potentially applicable to climate change risk disclosure and the impact of this guidance are discussed below.

b. Current Securities Laws, Regulations, and Guidance Requiring Disclosure of Climate-Change Issues

The disclosure obligations imposed by the Securities Act of 1933[225] and the Securities Exchange Act of 1934[226] and implementing regulations are designed to provide for full and fair disclosure of material information that would enable investors to make informed investment and voting decisions and to prevent disclosures from being misleading. Federal securities laws require periodic reporting and reporting for new offerings. Directors and officers can be liable for material misrepresentations or omissions.[227] SEC Rule 10b-5 prohibits both misleading disclosure as well as the omission of facts necessary to make statements to investors not misleading.[228]

SEC forms and regulations do not contain disclosure requirements specific to climate change-related risk. Publicly traded companies, however, have an obligation

---

224. Commission Guidance Regarding Disclosure Related to Climate Change, 75 Fed. Reg. 6290, 6295 (Feb. 8, 2010).
225. 15 U.S.C. §§ 77a, *et seq.*
226. *Id.* §§ 78a, *et seq.*
227. 17 C.F.R. § 240.10b-5.
228. *Id.* In addition, state securities laws (such as California's Blue Sky Laws) and common-law theories such as fraud and negligent misrepresentation could form the basis of liability.

to disclose certain kinds of environmental risks under existing SEC Regulation S-K, which governs the contents of periodic reports filed with the SEC. The SEC 2010 Guidance identified which of the existing environmental disclosure requirements could apply in the climate change context and examples of climate change issues that a company may need to consider in preparing disclosures.[229]

Regulation S-K, Item 101 may require climate change-related reporting. Item 101 requires companies to disclose material effects that compliance with federal, state and local provisions "regulating the discharge of materials into the environment, or otherwise relating to the protection of the environment, may have upon the capital expenditures, earnings, and competitive position of the company."[230] The SEC stated in the 2010 SEC Guidance that Item 101 may require specific disclosure of climate change-related factors should the cost required for environmental control facilities rise to the level of material capital corporate expenditures.[231]

Regulation S-K, Item 503, requires discussion of significant risk factors faced by publicly traded companies.[232] These risk factors often include discussion of environmental risks when environmental issues are significant enough to trigger disclosure under Item 101. Specifically, the SEC states that Item 503(c) may require disclosure of climate change-related risk factors in the context of "existing or pending legislation or regulation."[233] As EPA expands Clean Air Act requirements related to GHGs, companies may have reporting obligations under these items.

Under Regulation S-K, Item 103, companies are required to disclose "material pending legal proceedings other than ordinary routine litigation incidental to a company's business."[234] Companies subject to this provision often have to disclose material environmental proceedings. This is because the requirement specifies that administrative or judicial proceedings arising under provisions relating to the discharge of materials into the environment or for the primary purpose of protecting the environment are not considered ordinary and routine. The 2010 SEC Guidance suggests companies likely would have to report climate change-related litigation pursuant to this provision.

Regulation S-K, Item 303, Management's Discussion and Analysis of Financial Condition and Results of Operations (MD&A), requires companies to "[d]escribe any known trends or uncertainties that have had or that the registrant reasonably

---

229. Commission Guidance Regarding Disclosure Related to Climate Change, 75 Fed. Reg. 6290, 6295 (Feb. 8, 2010) (to be codified at 17 C.F.R. pts. 211, 231, 241).
230. 17 C.F.R. § 229.101(c)(xii).
231. Commission Guidance Regarding Disclosure Related to Climate Change, 75 Fed. Reg. at 6295.
232. 17 C.F.R. § 229.503.
233. Commission Guidance Regarding Disclosure Related to Climate Change, 75 Fed. Reg. at 6296.
234. 17 C.F.R. § 229.103.

expects will have a material favorable or unfavorable impact on net sales or revenues or income from continuing operations."[235] In considering the MD&A requirement, the 2010 SEC Guidance states that management must first evaluate "whether the pending legislation or regulation is reasonably likely to be enacted."[236] If management finds that the legislation is likely to pass and "have a material effect on the registrant," then MD&A disclosure is required.[237]

The SEC 2010 Guidance also suggested that traded entities whose businesses are reasonably likely to be affected by international agreements related to emissions trading or climate change should consider potential impacts in satisfying their disclosure obligations based on the MD&A and materiality principles discussed above.[238] The SEC 2010 Guidance also states that businesses should consider how significant physical effects of climate change could affect their operations and results.[239] Finally, pursuant to Item 101, companies may need to disclose where they shift operations to take advantage of new opportunities that arise out to climate change.[240]

Although not discussed in the 2010 SEC Guidance, the Sarbanes-Oxley Act requirements also could be relevant to climate change risks if such risks are deemed "material." Section 302 of the Sarbanes-Oxley Act requires CEOs and CFOs to certify that a financial report "does not contain any untrue statement of material fact or omit to state a material fact necessary in order to make the statements made, in light of the circumstances under which such statements were made, not misleading."[241]

In addition, the SEC requires companies to file financial statements in their annual and quarterly reports that are prepared in accordance with U.S. Generally Accepted Accounting Principles (GAAP). Pursuant to GAAP Statement of Financial Accounting Standards No. 5, Accounting for Contingencies, companies must recognize an environmental loss contingency in its financial statements if a loss is probable and the amount of the expected loss is material and reasonably estimable.

4. Climate Change-Related Activity in New York Pursuant to the Martin Act

Some states have chosen to move ahead without waiting for SEC guidance. New York has this option because of its unique authority under its Martin Act of 1921.[242] The Martin Act gives the New York attorney general power to investigate securities

---

235. Id. § 229.303.
236. Commission Guidance Regarding Disclosure Related to Climate Change, 75 Fed. Reg. at 6296.
237. Id.
238. Id.
239. Id.
240. Id.
241. 15 U.S.C. § 7241(a)(2).
242. N.Y. Gen. Bus. Law §§ 352–53 (McKinney 2011); N.Y. Exec. Law § 63(12) (McKinney 2012).

transactions and seek injunctive relief.²⁴³ In 1955, the New York legislature added section 352-c to the Martin Act, giving the attorney general the power to seek criminal indictments in securities fraud cases.²⁴⁴ New York's Executive Law section 63(12) further expands the Attorney General's power by authorizing him to bring special proceedings against any person or business committing repeated or persistent fraudulent or illegal acts. Essentially, the Martin Act prohibits the same type of fraud as federal securities laws, except that the Martin Act does not require a showing of scienter or proof that the subject misrepresentation was made knowingly.²⁴⁵ Consequently, under the Martin Act, individuals and businesses can be found guilty of securities crimes without knowing that their actions violated any law.

Former New York Attorney General Andrew Cuomo used these laws to pursue power companies for failing to make adequate climate change-related disclosures to the SEC. In September 2007, Cuomo subpoenaed the executives of several high carbon-emitting energy companies, including Dynegy, Xcel, AES, Dominion Resources, and Peabody Energy, seeking information on whether disclosures to investors in SEC filings adequately described the companies' climate change-related risks. Cuomo ultimately reached settlements with Xcel, Dynegy, and AES, wherein the three companies agreed to analyze material financial risks associated with climate change in their future SEC filings.²⁴⁶ Additionally, Xcel, Dynegy, and AES committed to disclose the following in future filings:

> (1) material financial risks associated with present and future regulations of GHGs, litigation, and physical impacts of climate change;
> (2) to the extent current GHG emissions materially affect financial exposure from climate change risk;
>> (a) estimated GHG emissions;
>> (b) projected increases in GHG emissions from planned coal-fired power plants;
>> (c) company strategies for reducing, offsetting, limiting, or otherwise managing its global warming pollution emissions and expected global warming emissions reductions from these actions; and
>
> (3) corporate governance actions related to climate change, including if environmental performance is incorporated into officer compensation.²⁴⁷

---

243. GEN. BUS. § 352–53.
244. GEN. BUS. §§ 352-c; *see also* People v. Landes, 645 N.E.2d 716 (N.Y. 1994).
245. *See* People v. Federated Radio Corp., 154 N.E. 655, 658 (N.Y. 1926).
246. *In re* Dynegy, AOD No. 08-132, http://www.oag.state.ny.us/sites/default/files/pdfs/bureaus/environmental/Attachment%20E-1.pdf; *In re* Xcel, AOD No. 08-012, http://www.oag.state.ny.us/sites/default/files/pdfs/bureaus/environmental/Attachment%20E%20--%20Xcel%20AOD.pdf; *In re* AES Corp., AOD No. 09-159, http://www.ag.ny.gov/sites/default/files/press-releases/archived/AES%20AOD%20Final%20fully%20executed.pdf.
247. *Id.*

Through use of state law, the former New York attorney general pushed the envelope for climate change risk disclosure by companies trading in New York. Given this initial success with top energy companies, New York eventually may broaden its pursuit of more-complete climate change disclosures to other carbon-intensive companies.

5. Voluntary Climate Change Disclosure Programs

Given the uncertain applicability of SEC regulations to climate change risk disclosure, voluntary disclosure programs have developed to address the needs and demands of investors. The main voluntary programs are the Carbon Disclosure Project, Principles of Responsible Investment, the Dow Jones Sustainability Index, and the FTSE4Good Index. Participating companies sometimes have the option to keep their disclosures confidential (so that the responses are only reviewed by the receiving organization),[248] but there is increasing public pressure for companies to make their full disclosures public and to provide greater and greater disclosures each year through these voluntary programs. In addition to voluntary disclosure programs, some companies also voluntarily disclose climate change-related information in annual reports and other public reports. All of these disclosures are significant, even though they are voluntary, because plaintiffs could use the information disclosed as the basis for their claims. This has ramifications for both potential claims liability and potential enterprise liability because many insureds and insurers participate in these programs, and include information about climate change in public reports.

> Voluntary climate change disclosure programs create risk of both potential claims liability and potential enterprise liability, because plaintiffs may use the voluntarily disclosed information as a basis for claims.

a. Carbon Disclosure Project

One of the best-known voluntary disclosure programs is the Carbon Disclosure Project (CDP). The CDP system is external to the securities regulation system, and companies may choose whether to make their responses publicly available.

Participation in the CDP among leading global, European, and North American companies has continued to show higher response and public disclosure rates, in part due to peer pressure and investor demands, as a review of available data from 2008

---

248. Nevertheless, the responses would be discoverable in litigation.

and 2011 reflect. In 2008, 77 percent of companies in the Global 500 responded to the CDP, which is "consistent" with the level achieved in 2007.[249] Of the 383 companies responding, 58 Global 500 companies completed the CDP for the first time in 2008.[250] European and North American companies held the highest response rates within the Global 500, with 83 percent and 82 percent response rates, respectively, compared with only 50 percent of Asian Global 500 companies.[251] An overwhelming majority of companies responding to the CDP chose to make their responses publicly available in 2008, and 85 percent of the Global 500 companies publicly disclosed their responses in 2008.[252] Similarly, 77 percent of S&P 500 and 68 percent of Financial Time Stock Exchange (FTSE) 350 responding companies publicly disclosed their 2008 CDP responses.[253]

By 2011, participation in the CDP had continued trending upward, with 81 percent of companies in the Global 500 providing responses.[254] As in prior reporting years, most of the companies responding (in this case, about 89 percent) chose to make their responses publicly available.[255] And this time around, 86 percent of the responding S&P 500 and 79 percent of FTSE 350 responding companies publicly disclosed their 2011 CDP responses.[256]

These rates and their progression during the past few years demonstrate an increased willingness among leading companies throughout the world to respond and allow publication of their CDP responses. Companies choosing not to publish their responses are now in the minority among the Global 500, FTSE 350, and the S&P 500 reporting companies.

CDP also is involved in standard setting for climate change disclosure. The CDP acts as the secretariat to, and advancing the causes of, the Climate Disclosure Standards Board (CDSB), which was formed in 2007 at the annual meeting of the World

---

249. *See* CARBON DISCLOSURE PROJECT, CDP GLOBAL 500 REPORT 2008, *available at* http://search.cdproject.net/reports.asp.
250. *Id.*
251. *Id.*
252. *See* CARBON DISCLOSURE PROJECT, CDP 2008 QUICK FACTS, *available at* http://search.cdproject.net/reports.asp.
253. *Id.*
254. *See* CARBON DISCLOSURE PROJECT, CDP GLOBAL 500 REPORT 2011: ACCELERATING LOW CARBON GROWTH, *available at* https://www.cdproject.net/CDPResults/CDP-G500-2011-Report.pdf.
255. *Id.*
256. *See* CARBON DISCLOSURE PROJECT, S&P 500 REPORT 2011, *available at* https://www.cdproject.net/CDPResults/CDP-2011-SP500.pdf; CARBON DISCLOSURE PROJECT, CDP FTSE 350 REPORT 2011: CAN UK PLC HELP MEET THE CARBON BUDGETS?, *available at* https://www.cdproject.net/CDPResults/CDP-2011-FTSE350.pdf.

Economic Forum.[257] CDSB's mission is to develop a globally accepted framework, based on existing standards, for corporate reporting on climate change.[258] On May 25, 2009, CDSB officially launched for comment its framework at the World Business Summit on Climate Change in Copenhagen, Denmark.[259]

The CDSB released its first edition of this framework in September 2010.[260] In this edition, the CDSB relied on three sources:

[1] the qualitative characteristics and constraints of decision-useful financial reporting information set out in the IASB's May 2008 Exposure Draft of An Improved Conceptual Framework for Financial Reporting;

[2] the Greenhouse Gas Protocol: A Corporate Accounting and Reporting Standard (Revised Edition) developed by the World Resources Institute and World Business Council for Sustainable Development associated with regional program protocols; and

[3] the International Organization for Standardization's ISO 14064-1—specification with guidance at the organizational level for quantification and reporting of greenhouse gas emissions and removals.[261]

The resulting framework addresses three aspects of climate change-related disclosure, including the determination of what disclosures should be made, the preparation of those disclosures, and the presentation to investors.[262]

b. Principles for Responsible Investment

The United Nations–backed Principles for Responsible Investment (PRI) Initiative is a network of international investors working together to put six principles for responsible investment into practice. The principles are:

1. We will incorporate ESG issues into investment analysis and decision-making processes.
2. We will be active owners and incorporate ESG issues into our ownership policies and practices.
3. We will seek appropriate disclosure on ESG issues by the entities in which we invest.

---

257. CDSB, http://www.cdsb-net.
258. *Id.*
259. Press Release, CDSB, Groundbreaking Proposals Unveiled for the Inclusion of Climate Change Data in Annual Reports (May 25, 2009), http://www.cdsb.net/file/36/cdsb_press_release.pdf.
260. CDSB, CLIMATE CHANGE REPORTING FRAMEWORK—EDITION 1.0 (Sept. 2010), *available at* http://www.cdsb.net/file/8/cdsb_climate_change_reporting_framework_2.pdf.
261. *Id.* at 14.
262. *Id.* at 7.

4. We will promote acceptance and implementation of the Principles within the investment industry.
5. We will work together to enhance our effectiveness in implementing the Principles.
6. We will each report on our activities and progress towards implementing the Principles.[263]

The program focuses on environmental, social, and corporate governance (ESG), which can include climate change considerations in investment and governance. The initiative began in 2005, and as of April 2012, more than 1,000 investment institutions have become signatories, with assets under management approximately $30 trillion.[264]

Signatories must report to PRI annually. The reporting process is mandatory and helps the PRI to evaluate signatories' progress in implementing the six principles.[265] As with any of the other voluntary programs necessitating disclosure once a party becomes a signatory, disclosures can have legal or reputational consequences.

c. Dow Jones Sustainability Index

The Dow Jones Sustainability Indexes (DJSIs) are premised on the integrated assessment of economic, environmental, and social criteria with a strong focus on long-term shareholder value.[266] "The indexes serve as benchmarks for investors who integrate sustainability considerations into their portfolios, and provide an effective engagement platform for companies who want to adopt sustainable best practices."[267]

Launched in 1999, the DJSIs were the first global indexes tracking the financial performance of leading sustainability-driven companies worldwide.[268] The DJSIs use a defined set of criteria to assess the opportunities and risks for eligible companies. A major source of information for the DJSIs is the questionnaire, developed by Strategic Asset Management USA, Inc. (SAM). Companies participating in the annual review complete this questionnaire. Then, SAM's sustainability analysts identify specific challenges for particular DJSI sectors and subsequently select criteria that enable them to identify the leading companies in terms of economic, environmental, and social issues.

---

263. *Principles*, PRI, http://www.unpri.org/principles/.
264. *About Us*, PRI, http://www.unpri.org/about/.
265. *Frequently Asked Questions*, PRI, http://www.unpri.org/faqs/.
266. *See* http://www.sustainability-indexes.com/.
267. *Index Family Overview*, DJSI, http://www.sustainability-indexes.com/dow-jones-sustainability-indexes/index.jsp.
268. *Id.*

Because climate change strategy is embedded within the environmental assessment of companies, that strategy directly influences companies' DJSI performance and thus investor perception of the companies as industry leaders. Companies are ranked within their industry group based on the SAM assessment and selected for the DJSIs if they are among the sustainability leaders in their field. Inclusion in the DJSIs yields several tangible and intangible benefits including: (1) public recognition as an industry leader in key strategic areas; (2) stakeholder recognition including legislators, customers, and employees; (3) increased visibility from publication of the index and entitlement to use the official "Member of DJSI" label; and (4) increased financial benefit resulting from eligibility to be included in DJSI-based portfolios.

d. FTSE4Good Index Series

Launched in 2001, the FTSE4Good Index Series (FTSE4Good) measures the corporate responsibility performance of companies in an effort to facilitate investment in those companies.[269] Similar to the DJSIs, the FTSE4Good considers climate change activities, among other things, in assessing companies' inclusion within the index. Unlike the DJSIs, however, the FTSE4Good has issued specific criteria for evaluating companies' climate change strategy in an effort to force transparency into the management of the series' criteria.

The series consists of five benchmark indices covering the global and European regions, including the United States, Japan, and the United Kingdom.[270] By using the series as an investment universe, investors can ensure that their portfolios will evolve to meet new environmental challenges. To be included in the indices, companies must establish that they are working toward climate change mitigation and adaptation, along with other goals. FTSE4Good criteria typically are based on international standards. But because there is no global consensus for measuring climate change efforts, the series has reached out to a broad range of stakeholders, including NGOs, governmental bodies, consultants, academics, the investment community, and the corporate sector in order to develop its standards.[271]

The "Key Principles and Challenges" of the FTSE4Good Climate Change Criteria include an evaluation of companies' climate change:[272]

---

269. *FTSE4Good Index Series*, FTSE, http://www.ftse.com/Indices/FTSE4Good_Index_Series/index.jsp (last visited Feb. 1, 2012).

270. *See* FTSE, FTSE4Good Index Series: Fact Sheet (2008), http://www.ftse.com/Indices/FTSE4Good_Index_Series/Downloads/FTSE4Good_Factsheet.pdf.

271. Id.

272. *FTSE4Good Climate Change Criteria*, FTSE, at 1, http://www.ftse.com/Indices/FTSE4Good_Index_Series/Downloads/FTSE4Good_Climate_Change_Criteria.pdf (last visited Feb. 1, 2012).

- *Policy*: contribution to scientific understanding on climate change and participation in public policy frameworks addressing climate risk;
- *Management*: establishment of systems to effectively manage climate risk through use of targets;
- *Disclosure*: revelation of GHG emissions using a standardized methodology for compiling GHG data;[273]
- *Performance*: reductions in absolute GHG emissions (adjusted for changes in company structure) over time; and
- *Scope*: applies only to companies' own operational GHG emissions and product GHG emissions.[274]

These criteria recognize that companies have an important role to play in developing effective climate change regulation, and companies are encouraged to play a constructive part in the public policy process.[275] These FTSE4Good criteria also may take into account any credible evidence that a company has deliberately and consistently misrepresented the scientific consensus on climate change (as represented by the IPCC reports), or attempted to undermine public policy frameworks that aim to reduce GHG emissions (including, but not limited to, those regarding mandatory emission reductions).[276] FTSE4Good has indicated its intent to implement more demanding climate change criteria over time, but in the meantime recognizes that "a very substantial number of companies will have difficulty in meeting the current criteria."[277]

As the criteria used by the series are evolving, the Responsible Investment Unit at FTSE is working with companies affected by the introduction of new criteria in an effort to help them work toward meeting the series' corporate responsibility standards. More than 200 companies globally have responded to FTSE4Good's more stringent environmental criteria to improve their practices. FTSE4Good has deleted 85 of them from the series due to their failure to meet the challenging criteria. These results exemplify the growing challenge for companies to step up the pace of their climate change activities in order to meet more exacting investor demands.

---

273. FTSE may allow some flexibility in the early stages of criteria implementation, reflecting that no single methodology is yet accepted as the single global standard. *Id.*

274. Upstream emissions (from suppliers' activities and/or the production of raw materials) may be included as the criteria continues to evolve. *Id.*

275. *Id.* at 2.

276. *Id.*

277. *Id.* at 4.

e. The Climate Registry

The Climate Registry is a not-for-profit organization formed by a coalition of U.S. states, Mexican states, Canadian provinces, and Native American tribes that sets "consistent and transparent standards to calculate, verify and publicly report greenhouse gas emissions (GHG) into a single registry."[278] It emerged out of the California Climate Action Registry. Its General Reporting Protocol, issued on March 31, 2008, is based on the protocol published by the World Resources Institute (WRI) and World Business Council for Sustainable Development (WBCSD).[279] The Climate Registry provides companies with a consistent set of standards and a centralized electronic reporting platform for reporting GHG emissions throughout North America.[280] It is the goal of the Registry that when the individual member states, provinces, and nations mandate reporting, the reporting companies will be authorized to use the Registry's protocol and database to report emissions.[281] Although reporting is voluntary at this time, member companies include Alcoa, PG&E, and Shell Oil Co., among others.[282] All reporting is public.[283]

The Climate Registry's reporting protocol requires that member companies report the six GHGs regulated by the Kyoto Protocol (CO_2, CH_4, N_2O, HFCs, PFCs, and SF_6).[284] Each member company must report both their direct (Scope 1) and indirect (Scope 2) emissions from each facility, instead of from its entire organization as the WRI and WBCSD standards require.[285] Although not required, many companies also report Scope 3 emissions, which are a "consequence of a company's activities, but

---

278. Brochure, The Climate Registry, http://the climateregistry.org/downloads/Registry_Brochure.pdf. The Climate Registry reports that all Canadian provinces, 39 U.S. states and the District of Columbia, six Mexican states, and three Native Indian nations are board members of the coalition. *See Board of Directors*, THE CLIMATE REGISTRY, http://www.theclimateregistry.org/about/board-of-directors (last visited Mar. 7, 2012).

279. *Climate Registry Announces Release of General Reporting Protocol for Greenhouse Gas Emissions*, 1 CLIMATE CHANGE L. & POL'Y REP., at 9 (May 2008) (citing THE GREENHOUSE GAS PROTOCOL: A CORPORATE ACCOUNTING AND REPORTING STANDARD (rev. ed. Mar. 2004)).

280. *Id.*

281. *Id.*

282. *Id.*; *see also List of Members*, THE CLIMATE REGISTRY, http://www.theclimateregistry.org/members (last visited Feb. 10, 2012).

283. *Public Reports*, THE CLIMATE REGISTRY, http://www.theclimateregistry.org/public-reports (last visited Feb. 10, 2012).

284. *Climate Registry Announces Release of General Reporting Protocol for Greenhouse Gas Emissions*, 1 CLIMATE CHANGE L. & POL'Y REP. 1, at 9 (May 2008).

285. *Id.* Reporting at the facility level is consistent with the framework established by the California Air Resources Board. Companies that are worried about the public nature of their reporting may request an exemption to report at the statewide level. *Id.* at 10. Direct emissions include those from on-site combustion, manufacturing processes, and from company-owned fleets. Indirect emissions are those associated with electricity and steam consumption. *See Frequently Asked Questions*, THE CLIMATE REGISTRY, http://www.theclimateregistry.org/about/faqs/ (last visited Mar. 12, 2012).

occur from sources not owned or controlled by [the company]."[286] A simple example of Scope 3 emissions would be those associated with employee business travel. The Climate Registry requires that all emissions results be verified.[287]

Now that EPA has finalized its GHG Reporting Rule, the future role of the Registry likely will change. Entities not subject to EPA rule may still elect to voluntarily report emissions to the Registry. The Registry also has requested that EPA partner with it to reduce the burden on companies facing multiple reporting requirements.[288] EPA reporting requirements probably will not supplant state GHG reporting until and unless state programs are specifically preempted. Washington State already has enacted its own GHG reporting rule. Regardless of how this plays out, plaintiffs will be looking at all available sources of emissions data to evaluate potential claims against emitters. If plaintiffs find discrepancies among a company's many emissions reports, this could also lead to mismanagement and concealment-related actions.

f. Periodic Stakeholder Reporting (Annual Reports)

In addition to SEC mandatory reporting and the voluntary reporting programs, companies frequently engage in periodic reporting to stakeholders regarding their operational performance and risk management. Just as with responses to mandatory SEC requirements and voluntary questionnaires from CDP and other organizations, climate change-related statements in annual reports could create liability for insureds. Already, in the *Kivalina* case (discussed in Section A.3), plaintiffs have cited to defendants' annual reports in support of their allegations.

F. Potential Litigation Arising out of Efforts to Address Climate Change Issues

1. Utilization of the National Environmental Policy Act (NEPA) and State Environmental Policy Acts (SEPAs) to Address Climate Change

a. National Environmental Policy Act (NEPA)

Plaintiffs have brought claims under the National Environmental Policy Act of 1969 (NEPA),[289] to address climate change issues. NEPA requires federal agencies to prepare environmental impact statements for "major federal actions significantly

---

286. *Climate Registry Announces Release of General Reporting Protocol for Greenhouse Gas Emissions*, 1 CLIMATE CHANGE L. & POL'Y REP., at 9 (May 2008).

287. *Frequently Asked Questions*, THE CLIMATE REGISTRY, http://www.theclimateregistry.org./about/faqs/ (last visited Mar. 12, 2012).

288. *See* US EPA, THE CLIMATE REGISTRY, http://www.theclimateregistry.org/government-services/mandatory-reporting/us-epa/ (last visited Feb. 2, 2012).

289. 42 U.S.C. § 4321.

affecting the quality of the human environment." Rather than establishing caps or setting standards, NEPA serves as an informational regulation, establishing processes for environmental impact review and mandatory disclosure of data to the public and decision-makers.

In the past decade, a number of NEPA lawsuits have arisen challenging the government for ignoring climate change during the environmental impact statement process. In one of the earliest successful lawsuits, in 2003, environmentalists forced the Surface Transportation Board to consider the climate effects of building a train line that would transport coal to Midwestern power plants.[290] In a 2008 case, plaintiffs forced the National Highway Traffic Safety Administration (NHTSA) to prepare an environmental impact statement analyzing the then proposed fuel economy standards for light trucks to include the monetized value of carbon emissions, among other things.[291] Courts also have used NEPA to require the examination of climate-related effects in projects federal agencies are supporting financially overseas. For example, in *Friends of the Earth, Inc. v. Mosbacher*, the court held that the Overseas Private Investment Corp. and Export Import Bank must examine the impact on the domestic environment of GHGs emitted by international fossil-fuel projects in accordance with NEPA.[292] Environmental groups continue to file climate change-related NEPA cases.[293]

Environmental groups also petitioned the Council on Environmental Quality (CEQ), which is charged with implementing NEPA, to issue more comprehensive guidelines targeted at the inclusion of climate-related impacts in the NEPA process. For example, on February 28, 2008, the International Center for Technology Assessment, the National Resource Defense Council, and the Sierra Club filed a formal legal petition with CEQ seeking amendment of NEPA regulations to clarify that climate change must be addressed in environmental reviews of federal projects and asking CEQ to issue a guidance document that will detail how to do so.[294]

CEQ has issued draft guidance on climate change and NEPA but nothing further. On October 8, 1997, CEQ issued a draft guidance policy that addressed the

---

290. Mid States Coal for Progress v. Surface Transp. Bd., 345 F.3d 520, 550 (8th Cir. 2003). On remand, the board used economic modeling software to analyze potential changes in low-sulfur coal consumption and to project air-quality effects, including the impact on U.S. CO_2 emissions. The EIS found inadequate in the *Mid States* case was supplemented with this analysis and held to be adequate in Mayo Found. v. Surface Transp. Bd., 472 F.3d 545, 553–54 (8th Cir. 2006).

291. Ctr. for Biological Diversity v. Nat'l Highway Traffic Safety Admin., 538 F.3d 1172 (9th Cir. 2008).

292. 488 F. Supp. 2d 889, 918 (N.D. Cal. 2007).

293. *See, e.g.*, Complaint for Declaratory Judgment and Injunctive Relief, WildEarth Guardians v. Salazar, No. 1:11-CV-00670 (D.D.C. Apr. 4, 2011), ECF No. 1; Complaint for Declaratory and Injunctive Relief, Mont. Envtl. Info. Ctr. v. BLM, No. 11-cv-00015-SEH (D. Mont. Feb. 7, 2011), ECF No. 1.

294. Int'l Ctr. for Tech. Assessment, et al., Petition Requesting that the Council on Environmental Quality Amend Its Regulations to Clarify that Climate Change Analyses Be Included in Environmental Review Documents (Feb. 28, 2008).

consideration of climate change under NEPA regulations.[295] On February 18, 2010, CEQ issued new draft guidance for public comment on when and how federal agencies must consider the impacts of proposed federal actions on climate change.[296] As of this writing, these guidance documents had not been finalized.

CEQ may eventually finalize guidance or issue binding regulations setting further standards for the assessment of GHG emissions and climate change in the environmental impact statement process. In the short term, however, individual project-based litigation, rather than federal regulatory activity, will probably continue to drive the forward momentum of the inclusion of climate effects in the NEPA process.

b. State Environmental Policy Acts (SEPAs)

States also have environmental policy laws that require state and local agencies to consider the likely environmental consequences of a proposal before approving or denying the proposal. California and other states are now using such laws to address GHG and climate change impacts. Environmental groups also are using such statutes to force state and local entities to consider GHGs and climate change when approving new projects.

(1) California Environmental Quality Act

> CEQA requires state agencies and private companies applying for environmental or construction permits to disclose and evaluate GHG emissions, as well as consider feasible mitigation measures.

California is using its California Environmental Quality Act (CEQA)[297] as a vehicle for achieving climate change-related regulatory goals. CEQA's environmental impact review requirements apply to public agencies and the activities or projects the agencies are responsible for approving and implementing if such activities have significant impacts on the environment. Thus, this process affects private stakeholders, such as corporations applying for environmental or construction permits. Modeled after NEPA, CEQA is similar to NEPA in its procedural requirements. CEQA, however, goes further than NEPA and other SEPAs in that it requires not just disclosure and evaluation of "significant" environmental impacts of proposed projects, but also the adoption of feasible measures to mitigate those impacts.

---

295. Draft Memorandum from Kathleen A. McGinty, Chairman, Council on Envtl. Quality, for Heads of Federal Agencies on Guidance Regarding Consideration of Global Climatic Change in Environmental Documents Prepared Pursuant to the National Environmental Policy Act (Oct. 8, 1997), *available at* http://www.mms.gov/eppd/compliance/reports/ceqmemo.pdf.

296. Council on Environmental Quality, National Environmental Policy Act (NEPA) Draft Guidance, "Consideration of the Effects of Climate Change and Greenhouse Gas Emissions," 75 Fed. Reg. 8,046 (Feb. 23, 2010).

297. Cal. Pub. Res. Code §§ 21000–21189.3 (West 2012).

California attorneys general have prepared numerous comment letters to agencies whose analysis under CEQA failed to properly analyze or mitigate a project's GHG emissions.[298] The California attorney general also has sued and entered into settlement with some of the entities for failing to consider and address GHG emissions.[299] California courts also have held that governmental entities must consider GHG emissions in the environmental impact review process.[300] Many groups continue to file climate change-related cases pursuant to CEQA.[301]

In order to clarify how GHGs and climate change must be addressed in the CEQA process, California passed a law, Senate Bill 97, requiring the Governor's Office of Planning and Research (OPR) to promulgate CEQA guidelines for mitigation of GHG emissions. As directed by S.B. 97, the Natural Resources Agency adopted Amendments to the CEQA Guidelines for GHG emissions on December 30, 2009.[302] On February 16, 2010, the Office of Administrative Law approved the amendments, and filed them with the secretary of state for inclusion in the California Code of Regulations.[303] The amendments took effect on March 18, 2010.[304] The amendments addressed identification of GHG emission sources, calculation and estimation of GHG emissions, consideration of cumulative effects, identification of mitigation measures and alternative methods, and identification of preferred mitigation strategies.[305]

Interest groups also have used CEQA in attempts to stop California from enacting climate change-related laws, like the California cap-and-trade program. Interest

---

298. *See* State of Cal. Office of the Attorney General, Comment Letters Filed Under the California Environmental Quality Act, *available at* http://loag.ca.gov/environment/ceqa/letters.

299. Settlements include: September 2007—Settlement with ConocoPhillips (proposed refinery expansion in Rodeo); December 2007—Settlement with Port of Los Angeles (GHG source tracking); March 2008—Settlement with Great Valley Ethanol (corn ethanol production plant); August 2008—Settlement with Cilion (corn ethanol production plant).

300. *See, e.g.*, Ctr. for Biological Diversity v. Town of Yucca Valley, No. CIVBS800607 (Cal. Super. Ct. San Bernadino County May 14, 2009) (ordering town to revise cumulative impact analysis to further address climate change in environmental impact review for proposed Walmart Supercenter); *see also* Ctr. for Biological Diversity v. City of Desert Hot Springs, No. RIC 464585 (Cal. Super. Ct. Riverside County Aug. 6, 2008); NRDC v. S. Coast Air Quality Mgmt. Dist., No. BS 110792 (Cal. Super. Ct. L.A. County July 28, 2008).

301. *See, e.g.*, Power Inn Alliance v. Cnty. of Sacramento Envtl. Mgmt. Dep't, No. C062994 (Cal. Ct. App. Mar. 15, 2011) (affirming denial of petition); Woodword Park Homeowners Ass'n v. City of Fresno, No. 09CECG00180 (Cal. Ct. App. Feb. 9, 2011) (affirming denial of petition); San Diego Navy Broadway Complex v. City of San Diego, 185 Cal. App. 4th 924 (Ct. App. 2010); Communities for a Better Env't v. City of Richmond, 184 Cal. App. 4th 70 (Ct. App. 2010).

302. Cal. Natural Res. Agency, *CEQA Guidelines*, http://ceres.ca.gov/ceqa/guidelines/ (last visited Aug. 23, 2012).

303. *Id.*

304. *Id.*

305. Cal. Code Regs. tit. 14, §§ 15064, 15064.7, 15065, 15086, 15093, 15125, 15126.2, 15126.4, 15130, 15150, 15183, 15064.4, 15183.5, 15364.5 (2012); *see also* Cal. Natural Res. Agency, *CEQA Guidelines Amendments* (Dec. 30, 2009), *available at* http://ceres.ca.gov/ceqa/docs/Adopted_and_Transmitted_Text_of_SB97_CEQA_Guidelines_Amendments.pdf.

groups argued that CARB was violating CEQA by taking steps to implement the cap-and-trade program without an adequate environmental impact review and consideration of alternatives.[306] A superior court agreed[307] but a state appellate court essentially lifted a stay pending further briefing on the issue in June 2011.

(2) SEPAs in Other States

Other states also are requiring that climate change-related issues be considered in the environmental impact review process. States including Massachusetts, New York, and Washington are applying their SEPAs in a manner that requires the consideration of climate change-related impacts. Other states with SEPAs may follow the California, Massachusetts, New York, and Washington trend and attempt to apply their SEPAs to climate change-related issues.

c. Implications

As demonstrated by the decision to halt a Walmart Supercenter project near Joshua Tree National Forest in California,[308] the environmental impact review process increasingly will require extensive coordination between developers and relevant public agencies on plans to consider, calculate, and mitigate GHG emissions related to any public project. The business of identifying GHG sources, calculating GHG emissions, and developing mitigation strategies is burgeoning because of this new SEPA regulation. Disputes, however, may arise between developers and engineers and other contractors responsible for calculating and achieving emissions reductions. Engineers and other contractors probably will look to professional liability insurance for defense and indemnification if any disputes arise. See Chapter 7.A.3.d, discussing implications for professional liability insurance.

Another potential consequence of SEPA regulation is delay. Developers will face additional costs associated with project delays resulting from failure to consider GHG issues and/or settlement payouts related to excessive GHG emissions. Although this is a business interruption of sorts, it probably cannot trigger business interruption or construction delay coverage because there is no physical loss. See Chapter 7.A.2.b, discussing implications for property (time element) coverage.

---

306. Petition for Writ of Mandate, Ass'n for Irritated Residents v. Cal. Air Res. Bd., No. CPF-09-509562 (Cal. Super. Ct. June 10, 2009).

307. Order Granting in Part and Denying in Part Petition for Writ of Mandate, Ass'n for Irritated Residents v. Cal. Air Res. Bd., No. CPF-09-503562 (Cal. Super. Ct. May 20, 2011).

308. Ctr. for Biological Diversity v. Town of Yucca Valley, No. CIVBS800607 (Cal. Super. Ct. San Bernadino County May 14, 2009) (ordering town to revise cumulative impact analysis to further address climate change in EIR for proposed Walmart Supercenter).

> Burgeoning SEPA regulation could lead to disputes among developers, engineers, and other contractors engaged in calculating GHG emissions and developing mitigation strategies. These professionals are likely to look to their professional liability policies for both defense and indemnification should disputes arise.

2. Disputes Arising out of Carbon and GHG Markets

Legislators in the United States have not passed comprehensive cap-and-trade legislation related to GHGs. Carbon-emissions trading, however, is ongoing in the European Union Emission Trading System (EU ETS). Both California and Quebec also have adopted GHG cap-and-trade systems that are likely to go into effect in 2013 unless they are delayed or subject to lengthy challenges. Because liabilities may arise out of emerging carbon and GHG markets, insurers and insureds should be aware of how carbon and GHG markets work and how they may lead to potential risks and claims.

Generally, pollution trading markets are based on a cap-and-trade system. The regulatory entity sets a cap and then issues a certain number of emissions allowances to regulated entities. At the end of a given compliance period, the regulated entity must turn in allowances equal to their emissions. If a regulated entity has fewer emissions than allowances, it can sell its additional allowances in the carbon or GHG market. If a regulated entity has more emissions than allowances, it enters the carbon or GHG market to either trade for more allowances or purchase offsets. "Carbon or GHG offsets are tradable financial instruments representing verified emission reductions by entities not subject to the governmental emissions cap."[309] Some cap-and-trade systems limit what portion of compliance obligations may be satisfied by offsets as opposed to allowances or emissions reductions by the regulated entity.[310]

---

309. Jonas Monast, Jon Anda & Tim Profeta, *U.S. Carbon Market Design: Regulating Emission Allowances as Financial Instruments*, at 11 (Duke Univ. Nicholas Inst. for Envtl. Policy Solutions, Working Paper No. CCPP 09-01, Feb. 2009), *available at* http://www.nicholas.duke.edu/ccpp/ccpp_pdfs/carbon_market_primer.pdf. *See also Environmental Commodities Trading*, J.P. MORGAN, http://www.jpmorgan.com/pages/jpmorgan/investbk/solutions/commodities/environmental (last visited Nov. 7, 2011). J.P. Morgan handles exchange-traded and over-the-counter (OTC) environmental products, including European Union Allowances (EUAs), Certified Emission Reductions (CERs) and Verified Emission Reductions (VERs).

310. *See generally* Directive 2003/87/EC of the European Parliament and of the Council of 13 October 2003 Establishing a Scheme for Greenhouse Gas Emission Allowance Trading Within the Community and Amending Council Directive 96/61/EC and later amendments, 2003 O.J. (L 275), *available at* http://eur-lex.europa.eu/LexUriServ/LexUriServ.do?uri=CONSLEG:2003L0087:20090625:EN:PDF; REGIONAL GREENHOUSE GAS INITIATIVE (RGGI), CO_2 *Offsets*, http://www.rggi.org/market/offsets.

Some systems also could permit various types of allowance derivatives, like futures, so regulated entities can manage the risk stemming from allowance shortages or price volatility.[311]

There are potential risks associated with carbon and GHG market systems. Regulated emitters (regulated entities engaged in trading for compliance purposes), carbon market investors,[312] GHG reduction and footprint auditing consultants, and low-carbon technology suppliers may face liability related to their roles in the carbon market. Some of those entities could face breach of contract claims, errors and omissions claims, or fines and penalties relating to their participation in the carbon market. They could also face shareholder actions if they cause or contribute to plummeting stock prices. In addition, carbon traders and regulated entities may face shareholder action if their investments turn out to be unwise or a missed opportunity in the market has a significant financial impact. Misstatements about carbon offsets to consumers also might lead to regulatory enforcement actions or lawsuits by consumers.[313]

There have been no major disputes or lawsuits with respect to carbon credits or carbon credit trading in the United States, largely because there currently is no nationwide, mandatory cap-and-trade system. Emerging environmental integrity problems in the EU ETS (the largest mandatory cap-and-trade program implemented to assist member countries in meeting their Kyoto Protocol obligations), however, illustrate the potential environmental integrity issues facing emerging U.S. and global carbon markets. For example, one project verifier, Det Norske Veritas (DNV),[314] was suspended by the United Nation's Clean Development Mechanism Executive Board for insufficient oversight and monitoring of offset projects, putting at risk significant investment by the carbon trading firm Ecosecurities for compliance grade offsets to be sold in the EU ETS.[315] The United Nations later reinstated DNV as a project verifier.[316] This example illustrates that providers in carbon markets may be exposed to breach of contract and professional negligence claims. Failures to properly monitor and verify carbon credits also may place regulated client entities who rely on such

---

311. Monast et al., *supra* note 309, at 7–8.

312. Carbon market investors are involved in either the trading of emission credits or direct investment in offset projects with a view to receiving a share of subsequently derived carbon credits. In the private sector, two leading-carbon market investors are Ecosecurities and Climate Change Capital.

313. See discussion *infra* Section F.3 (on greenwashing, including the FTC's new Green Guides).

314. DNV is a global provider of services for managing risk. In particular, DNV was the first verifier accredited under the Kyoto Protocol. Press Release, Det Norske Veritas, DNV Announces Major Expansion of Climate Change Services in North America, *available at* REUTERS, Mar. 30, 2009, http://www.reuters.com/article/pressRelease/idUS154744+30-Mar-2009+PRN20090330.

315. Michael Szabo, Analysis, *DNV Suspension Another Jab at Battered Carbon Dioxide Scheme*, REUTERS, Dec. 2, 2008, http://www.reuters.com/article/2008/12/02/us-carbon-dnv-idUSTRE4B04K120081202.

316. James Murray, *DNV Wins Back UN Authorisation for CDM Project Approval*, BUS. GREEN, Feb. 16, 2009, http://www.businessgreen.com/bg/news/1804681/dnv-wins-un-authorisation-cdm-project-approval.

credits at significant risk of being unable to meet their compliance obligations under mandatory cap-and-trade programs. Claims against companies like DNV by regulated entities and/or shareholders potentially could implicate both professional liability and D&O policies (subject to applicable fraud exclusions).

In addition to verification and accuracy problems, the EU ETS also has faced other controversies such as over-allocation of allowances, price volatility, and allegations that energy companies received windfall profits as a result of the initial trading.[317] U.S. systems could face similar issues. Comprehensive U.S. GHG legislation at some future date may attempt to address potential integrity issues in U.S. trading systems.

Carbon market participants also face risks associated with participation in an uncertain global regulatory framework. Global market participants face regulatory risks, including "those associated with host-country and international policies governing emissions-reduction projects such as project approval, validation, and verification" and "host-country investment and political risks that could alter climate change policies and obligations, such as host-country instability, expropriation of credits, contract frustration, credit confiscation, and more."[318]

Given the development of GHG and carbon markets and associated risks for various stakeholders, insurers may eventually face related claims and also opportunities to assist customers with risk management. With respect to existing policies, insurers may already control some of the potential risks through use of particular policy triggers and exclusions. Nonetheless, underwriters may want to consider how a carbon or GHG trading risk could alter existing risk profiling. Many insurers and other financial institutions also are working to develop products to insure offset delivery and lessen the impact of unexpected policy changes and market volatility.

3. Greenwashing Litigation

In response to consumer demand, companies have developed products that are more natural, green, and environmentally sensitive in recent years. Companies also have sought to design products that leave a smaller carbon footprint. As companies have responded to consumer demand for greener products and services by embarking on green marketing campaigns, they also have found themselves exposed to "greenwashing" claims. "Greenwashing" is not subject to one single uniform definition but generally it is understood to be "disinformation disseminated by an organisation so as

---

317. *See* A. Denny Ellerman & Paul L. Joskow, The European Union's Trading System in Perspective (Pew Ctr. Report on Global Climate Change May 2008).
318. Evan Mills & Eugene Lecomte, Ceres, From Risk to Opportunity, How Insurers Can Proactively and Profitably Manage Climate Change 23 (Aug. 2006).

to present an environmentally responsible public image."[319] "Greenwashing" lawsuits, including class actions, accusing corporations of marketing the environmental and eco-friendliness of their products or services in a false or misleading way have sprung up across the United States in recent years. Plaintiffs have filed greenwashing lawsuits involving claims related to everything from household cleaners to hybrid automobiles.

In addition to consumer actions, the FTC also has become involved in scrutinizing green marketing claims in an effort to protect consumers from misleading environmental advertising. In late 2010, FTC proposed revisions to its Green Guides,[320] and it has stepped up enforcement actions in recent years. Companies also have challenged their competitors' green assertions through the forum provided by the National Advertising Division of the Council of Better Business Bureaus.

Greenwashing litigation potentially could create a new source of insurance claims. False-advertising claims are sometimes covered under CGL policies, although the failure to conform exclusion may bar some kinds of greenwashing claims if the allegations are based on a company's own product versus disparaging another company's product. See Chapter 7.A.3.b for implications for CGL coverage. If greenwashing claims ever resulted in a material financial impact to a corporation, shareholder derivative suits and other actions by shareholders could potentially lead to new D&O exposures. See Chapter 7.A.3.a for implications for D&O coverage.

> As consumers continue to demand "green" products, greenwashing litigation is emerging.

a. Greenwashing and Consumer Actions Related to Misleading or False Advertising

Greenwashing lawsuits have been filed against companies in a variety of industries and trades, including construction companies, retailers, automakers, food manufacturers, cosmetic companies, and manufacturers of cleaning supplies. The vast majority of greenwashing cases are consumer product–related, false advertising actions that have arisen under state laws. Some have been filed as class actions. Some of the recent cases are summarized and analyzed below.

A number of cases have been filed related to hybrid vehicles. In one of the first cases, *Paduano v. American Honda Motor Co.*, a consumer sued Honda in California

---

319. CONCISE OXFORD ENGLISH DICTIONARY 624 (10th ed. 2002).
320. Press Release, Fed. Trade Comm'n, Federal Trade Commission Proposes Revised "Green Guides" (Oct. 6, 2010), http://www.ftc.gov/opa/2010/10/greenguide.shtm.

state court because he was dissatisfied with the gas mileage of his hybrid Honda vehicle.[321] In addition to asserting state and federal warranty claims, the plaintiff in that case alleged false advertising and deceptive practices relating to the car company's statements about fuel efficiency. In 2006, the trial court ruled that Honda was entitled to summary judgment and dismissed the plaintiff's claims on the ground that the state's deceptive advertising claims were preempted by federal law—the Energy Policy and Conservation Act.[322] The court concluded that Honda's advertising complied with federal regulations on fuel economy advertising disclosures. In January 2009, California's Fourth District Court of Appeal affirmed the portion of the preemption ruling on the breach of warranty claims, but reversed the lower court's decision on deceptive practices and misleading advertising, ruling that the plaintiff could go forward with those claims. In 2009, Honda settled the case by agreeing to pay Paduano $50,000.[323]

Based on the same allegations of misleading statements about the mileage of hybrid cars, additional plaintiffs filed class action lawsuits against Honda in California.[324] In *True v. American Honda Motor Co.*, for example, True alleged that the fuel economy estimates Honda advertised for the Honda Civic hybrid could not be achieved under normal driving conditions. The U.S. District Court of the Central District of California denied Honda's motion to dismiss on preemption and other grounds.[325] On August 27, 2009, the court preliminarily certified a settlement class, preliminarily approved the initial proposed settlement, and directed notice be given to the class members.[326] On February 26, 2010, however, the court declined to approve the final settlement. The court cited various concerns with the fairness of the settlement.[327] The case was placed on inactive stay.

Parties from various Honda Civic cases continued to work on a global settlement of the pending class actions against Honda related to the Civic hybrid.[328] A San

---

321. Paduano v. Am. Honda Motor Co., No. GIC 852441, 2006 WL 6331968 (Cal. Super. Ct. San Diego County Sept. 15, 2006), *aff'd in part, rev'd in part*, 169 Cal. App. 4th 1453 (2009).
322. *Id.* (citing 49 U.S.C. § 32919(a)).
323. Settlement Agreement, Paduano v. Am. Honda Motor Co., No. GIC 852441 (Cal. Super. Ct. San Diego County Dec. 3, 2009) (available as Exhibit 1 to Docket No. 154, True v. Am. Honda Motor Co., No. 5:07-cv-287-VAP-OP (C.D. Cal. Feb. 17, 2010)).
324. Class Action Compl., *True v. Am. Honda Motor Co.*, No. 5:07-cv-287-VAP-OP (C.D. Cal. Mar. 9, 2007); Compl. for Equitable & Monetary Relief, *Lockabey v. Am. Honda Motor Co.*, No. 37-2010-00087755-CU-BT-CTL (Cal. Super. Ct. San Diego Cnty. Mar. 15, 2010); Class Action Compl., *Gibble v. Am. Honda Motor Co.*, No. 2:10-cv-6148-VAP-OP (C.D. Cal. Aug. 18, 2010); Class Action Compl., *Stouch v Am. Honda Motor Co.*, No. 2:10-cv-6236-VAP-OP (C.D. Cal. Aug. 20, 2010); Am. Class Action Compl. for Equitable & Monetary Relief, *Thieben v. Am. Honda Motor Co.*, No. BC 441424 (Cal. Super. Ct. L.A. Cnty. Aug. 17, 2010).
325. True v. Am. Honda Motor Co., 520 F. Supp. 2d 1175 (C.D. Cal. 2007).
326. True v. Am. Honda Motor Co., 749 F. Supp. 2d 1052, 1059 (C.D. Cal. 2010).
327. *Id.* at 1066–77.
328. True v. Am. Honda Motor Co., No. 5:07-cv-287-VAP-OP; Lockabey v. Am. Honda Motor Co., No. 37-2010-00087755-CU-BT-CTL; Gibble v. Am. Honda Motor Co., No. 2:10-cv-6148-VAP-OP; Stouch v. Am.

Diego judge presiding over the *Lockabey v. American Honda Motor Co.* class action tentatively approved a global settlement in March 2012.[329] In settlement of the claims, Honda agreed to pay $100 to $200 to each owner and to provide a $1,000 credit toward a new car.[330]

One plaintiff also succeeded in making a greenwashing claim against Honda in small-claims court. Heather Peters was awarded $9,867 after alleging she was misled about the potential fuel economy of her Honda.[331] Peters had opted out of the proposed class action settlement in order to try to obtain a bigger award.[332] More than 1,700 car owners have opted out of the settlement.

In July 2009, plaintiffs filed a greenwashing lawsuit against Toyota in California, *Bernstein v. Toyota Motor Sales USA, Inc.*, which also addressed the performance of hybrid vehicles.[333] The case was dismissed in August 2009 just weeks after it was filed because the plaintiff failed to properly allege any basis for federal jurisdiction.

Plaintiffs have filed greenwashing actions against other types of manufacturers. In March 2009, a consumer filed a class action against SC Johnson & Son, Inc. in the U.S. District Court for the Northern District of California, alleging that the maker of Windex is misleading consumers about the "environmental safety and soundness" of the cleaning product.[334] In question was the company's use of its "Greenlist" trademark. In this case, *Koh v. SC Johnson & Son, Inc.*, the plaintiff claimed the "Greenlist" trademark was misleading because it is not a third-party endorsement of the product's greenworthiness, but is instead a mark created and owned by SC Johnson. The complaint consisted of eight causes of action, the first five of which were based on provisions of California's Business and Professions Code (Unfair Competition Law): unlawful business acts and practices; unfair business acts and practices; fraudulent business acts and practices; misleading and deceptive advertising; and untrue advertising. The plaintiff also asserted causes of action (1) for fraud, deceit, and misrepresentation; (2) for unjust enrichment; and (3) under California's Consumer Legal Remedies Act (CLRA) for injunctive relief. Also included in the complaint was a

---

Honda Motor Co., No. 2:10-cv-6236-VAP-OP; Thieben v. Am. Honda Motor Co., No. BC 441424.

329. Notice of Entry of Order Granting Final Approval of Class Action Settlement and Application for Payment of Attorneys' Fees, Costs and Class Representative Payments, Final Judgment and Order of Dismissal with Prejudice Denying Objectors' Motions, Lockabey v. Am. Honda Motor Co., No. 37-2010-87755-CU-BT-CTL (Mar. 16, 2012).

330. Notice of Entry of Order, *supra* note 329; *see also* Bill Callahan, *Honda Hybrid Gas Mileage Settlement Wins Final Approval*, BLOOMBERG, Mar. 16, 2012.

331. Linda Deutsch, *Woman Takes Honda to Small-Claims, Wins Big*, INS. J., Feb. 2, 2012, *available at* http://www.insurancejournal.com/news/west/2012/02/03/234115.htm.

332. *Id.*

333. Complaint for Violation of California's Business and Professional Code Sections, Bernstein v. Toyota Motor Sales USA, Inc., No. 09-cv-03472-MHP (N.D. Cal. July 28, 2009).

334. Complaint, Koh v. SC Johnson & Son, Inc., No. 09-cv-00927 (N.D. Cal. Mar. 2, 2009).

demand for compensatory and punitive damages, which are recoverable under the CLRA. SC Johnson settled the case in July 2011.[335]

In *Women's Voices for the Earth, Inc. v. Procter & Gamble Co.*, a group of environmental and public safety groups filed a New York state court action against Procter & Gamble Co., Colgate-Palmolive, Inc., and other manufacturers of household cleaners in an attempt to force them to reveal the chemical ingredients of their products.[336] The plaintiffs sought damages and a court order compelling defendants to comply with New York State's Department of Environmental Conservation disclosure regulations. The complaint made clear that the plaintiffs were endeavoring to determine whether the manufacturers of the household items—such as Tide and Ajax—were falsely promoting their products as eco-friendly. Indeed, plaintiffs' lead attorney stated in media interviews that "[w]e have a great deal of concern about companies misleading the public whether products are green.[337] This lawsuit is about greenwashing in that it can help uncover when a company is putting things out there that ultimately aren't really true."[338] The court ultimately dismissed the plaintiffs' petition finding that the plaintiffs had no private right of action.[339]

In *Hill v. Roll International Corp.*, a California court held in a 2011 decision that the Green Drop on Fiji bottled water is not deceptive.[340] Hill had alleged that the Green Drop represented that Fiji bottled water was environmentally superior to other waters and suggested the product was endorsed by an environmental organization. Claiming that the Fiji product was not in fact environmentally superior or endorsed by an organization, Hill filed a proposed class action on behalf of herself and other consumers of Fiji bottled water, naming Roll International Corp. and Fiji Water Co., and asserting violations of California's Unfair Competition Law,[341] False Advertising Law,[342] and Consumers Legal Remedies Act,[343] plus common law fraud and unjust enrichment. A California appeals court affirmed the trial court's dismissal of the case for failure to state a claim.[344] The appellate court agreed that the Green Drop was not deceptive.

---

335. Press Release, SC Johnson, SC Johnson Settles Case Involving Greenlist Labeling (July 8, 2011), http://www.scjohnson.com/en/press-room/press-releases/07-08-2011/SC-Johnson-Settles-Cases-Involving-Greenlist-Labeling.aspx.

336. Women's Voices for the Earth, Inc. v. Procter & Gamble Co., 906 N.Y.S.2d 721 (Sup. Ct. 2010).

337. Tresa Baldas, *Claims of 'Greenwashing' on the Rise*, NAT'L L.J., June 10, 2009, *available at* http://www.law.com/jsp/cc/PubArticleCC.jsp?id=1202431342143.

338. *Id.*

339. *Women's Voices for the Earth, Inc.*, 906 N.Y.S.2d 721.

340 Hill v. Roll Int'l Corp., 128 Cal. Rptr. 3d 109 (Cal. App. 1st 2011).

341. Bus. & Prof. Code §§ 17200 *et seq.*

342. §§ 17500 *et seq.*

343. Civ. Code §§ 1750 *et seq.*

344. *Id.*

The number and types of greenwashing cases could increase as both consumer demand for green products and scrutiny of green-product advertising increases. Of note is a 2010 study by TerraChoice, an environmental marketing agency, which found that the number of green products increased by 73 percent from 2009 to 2010.[345] TerraChoice evaluated more than 5,000 products with green claims.[346] The TerraChoice study concluded that 95 percent of these products exaggerated their greenworthiness.[347]

Demand for green products decreased during the recession due to their generally higher cost, but demand is likely to pick up again with an improving economy. In addition to the types of claims filed to date, plaintiffs will probably file additional lawsuits related to products, including suits alleging misleading and false advertising based on a company's marketing of its reduced carbon footprint, GHG emissions, or reduced environmental impacts.

Greenwashing claims may have implications for CGL coverage. See Chapter 7.A.3.b. In order to manage a potential rise in claims, portfolio risk managers for insurers might consider use of exclusions for quality or performance of goods (failure to conform), fines and penalty, material published with knowledge of falsity, and breach of contract, as well as specific exclusions for green-related false advertising issues. In addition, underwriters should consider that some policies only grant defense and indemnity coverage for actions seeking "damages." These various provisions should be considered, along with potential coverage extensions, as greenwashing litigation evolves and claims increase.

Recent coverage decisions provide some insight into how greenwashing claims may be handled. For example, although it did not involve a greenwashing claim *per se*, the North Carolina Supreme Court recently held that coverage for claims regarding alleged false statements about an insured's *own* products are barred by the quality or performance of goods (failure to conform) exclusion in a CGL policy.[348] The coverage dispute in *Harleysville Mutual Insurance Co. v. Buzz Off Insect Shield, LLC* arose out of an underlying lawsuit by SC Johnson & Son, Inc. against its competitors Buzz Off Insect Shield, LLC and International Garment Technologies, LLC. SC Johnson alleged that the defendants falsely advertised the attributes of their insect-repellent clothing, and that this harmed SC Johnson. The question before the North Carolina

---

345. Press Release, TerraChoice, TerraChoice 2010 Sins of Greenwashing Study Finds Misleading Green Claims on 95 percent of Home and Family Products (Oct. 26, 2010), http://www.terrachoice.com/files/TerraChoice%202010%20Sins%20of%20Greenwashing%20Release%20-%20Oct%2026%202010%20-%20ENG.pdf.
346. Id.
347. Id.
348. Harleysville Mut. Ins. Co. v. Buzz Off Insect Shield, LLC, 692 S.E.2d 605 (N.C. 2010).

court was whether International Garment Technologies' CGL insurance carriers were required to defend it against SC Johnson's claims.[349] The court held that because SC Johnson only alleged it was injured by false statements defendants made about their *own* products (as opposed to SC Johnson's), the CGL policies' failure to conform exclusion dictated that there was no insurance coverage for SC Johnson's injury.[350] Thus, the insurance carriers were not required to defend their insured against SC Johnson's claims in that case. Coverage cases based on underlying greenwashing claims also may focus on the nature of the false advertising and disparagement allegations.

The availability of insurance coverage also may depend in part on the type of relief being sought by the plaintiff. In the greenwashing cases to date, plaintiffs have sought both damages and injunctive relief. In considering these cases—several of which have arisen under California law—it should be noted that damages (compensatory or punitive) are not recoverable for claims pursuant to that state's Unfair Competition Law, California Business and Professions Code § 17200, *et seq.*[351] There are basically three types of remedies available: injunctive relief, restitution, and civil penalties. Injunctive relief and restitution are available in both private-party and government actions. Civil penalties are available only in government enforcement actions. Courts have broad powers to fashion creative awards of injunctive relief. For example, in one false-advertising case, a California court exercised its injunctive power to require a ten-year mandatory disclosure in the form of a warning on the defendant's future products.[352] Outside of the greenwashing context, California courts have held that there is no coverage under a CGL policy for a section 17200 claim for damage due to advertising injury because (1) damages were not available under section 17200 and (2) unfair competition under the policy referred to common law but not statutory claims.[353]

b. Federal Trade Commission Focus on "Green" Marketing

In addition to consumer actions, the FTC has also increased its scrutiny of green marketing. In June 2009, the FTC filed suit against Kmart, Tender Corp., and Dyna-E International for making false and unsubstantiated claims that their products were biodegradable. The FTC alleged that promoting the products as green did not conform with the FTC's Guides for the Use of Environmental Marketing Claims ("Green

---

349. *Id.* at 608.
350. *Id.*
351. *See, e.g.*, Korea Supply Co. v. Lockheed Martin Corp., 29 Cal. 4th 1134, 1150–51 (Cal. 2003).
352. Consumers Union of United States, Inc. v. Alta-Dena Certified Dairy, 4 Cal. App. 4th 963, 972–74 (Ct. App. 1992) (requiring warning to be placed on all of dairy company's advertisements and products for the next ten years).
353. Bank of the West v. Superior Court, 2 Cal. 4th 1254 (1992).

Guides"), a set of regulations used by the FTC to determine whether a company's green marketing constitutes false or misleading use of environmental terms in product advertising. The Green Guides themselves are not enforceable, but if a corporation makes claims that are inconsistent with them, the FTC can take legal action under section 5 of the FTC Act, which prohibits unfair or deceptive practices. The FTC settled all three actions in July and August 2009; all three companies are barred from making deceptive "biodegradable" product claims and are required to support all other environmental product claims with "competent and reliable evidence."[354]

The FTC—which issued its first Green Guides in 1992—issued revised Green Guides in 2012.[355] The revised Green Guides address new categories like "renewable materials," "renewable energy," and "carbon offsets" and add specificity to guidance on general environmental benefit claims, environmental seals, and claims like "compostable," "degradable," and "recyclable." The Green Guides do not carry the force of law but they could create a standard of care that will apply in greenwashing cases. The FTC may consider that acts in violation of the Green Guides constitute a violation of the FTC Act.[356] See also Chapter 6.F.3.b.

4. Green Building Litigation

> Green building poses unique opportunities to develop and market new products. As green building becomes more prevalent, however, disputes arising from green building services will also increase, leading to the assertion of more insurance claims.

According to the EPA, green building is the practice of creating healthier and more resource-efficient models of construction, renovation, operation, maintenance, and demolition.[357] The underlying feature of green building is the efficient use of energy and water, which is usually achieved by using energy-efficient materials during the construction of the project and the integration of water and energy conservation systems. Green buildings also tend to incorporate eco-friendly concepts and materials into the use of the surrounding land and landscaping. The perceived benefits of green building

---

354. *In re* Dyna-E Int'l, Inc., FTC File No. 082 3187; *In re* Kmart Corp., FTC File No. 082 3186; *In re* Tender Corp., FTC File No. 082 3188; Press Release, Fed. Trade Comm'n, FTC Announces Actions Against Kmart, Tender an Dyna-E Alleging Deceptive 'Biodegradable' Claims (June 9, 2009), *available at* http://www.ftc.gov/opa/2009/06/kmart.shtm.
355. Press Release, FTC Issues Revised Green Guides, Oct. 1, 2012, http://www.ftc.gov/opa/2012/10/greenguides.shtm.
356. 16 C.F.R. § 260.1.
357. EPA, Green Building, http://www/epa.gov/greenbuilding/index.htm (last visited Oct. 2, 2012).

are reduced operating costs, recognition by consumers of environmental-friendliness, and the potential availability of government incentives, including tax benefits.

Green building indirectly relates to climate-change exposure because episodic climatic events may lead to destruction of property that is then replaced by green buildings and developers of new properties may look to green building to reduce GHG emissions. Thus, the green building movement presents possible new exposures as well as new opportunities for insurers. On the exposure side, green building may lead to disputes resulting in claims against CGL, professional, and construction-related policies. Green building could also result in new types of exposures to property insurance lines due to use of green roofs, water collection systems, renewable energy systems, and less familiar construction materials. On the flip side, green building also could drive demand for new services and insurance products, such as those that offer incentives for replacement with green materials.

This section reviews the background on green building standards, litigation that may result from increasing regulation and legislation, and potential contractual disputes involving green building.

a. Green Building Standards

Green building standards are emerging. In 1998, the U.S. Green Building Council (USGBC) developed the LEED Green Building Rating System. The LEED rating system evaluates the design, construction, and operation of newly constructed or renovated buildings and serves as the standard for sustainable buildings. LEED is a voluntary program but some jurisdictions are adopting regulations that require certain new structures to be LEED certified.[358] The federal government also has green building requirements for major construction projects, although appropriations bills have at times attempted to limit funding for obtaining LEED Silver and Gold certifications in certain departments.[359] Compliance with these various standards may lead to litigation.

The LEED requirement that certified buildings meet continued energy performance and efficiency standards post-certification also may lead to additional disputes. As part of LEED Version 3 (v3), which was introduced in April 2009, buildings seeking LEED certification will be required to agree to submit operational performance data (energy-efficiency information) on a recurring basis as a precondition to

---

358. D.C. Code § 6-1451.03 (2012) (setting green building requirements for large commercial projects).
359. *See* National Defense Authorization Act for Fiscal Year 2012, § 2820(b), P.L. 112-81, 125 Stat. 1695–96 (prohibiting use of FY2012 Department of Defense funds for achieving LEED platinum or gold certifications). This law also requires the Department of Defense to complete a cost-benefit analysis of green building policies.

certification.³⁶⁰ The LEED 2009 minimum program requirements for Existing Buildings Operations and Maintenance is that projects must commit to sharing with the USGBC all available actual whole-project energy and water usage data for a period of at least five years.³⁶¹

Shortly after the publication of LEED v3, it was announced that building projects can comply with the new performance requirements in one of three ways: (1) recertifying on a two-year cycle (using LEED for Existing Buildings: Operations and Maintenance), (2) providing utility usage data on an ongoing basis and annual basis, or (3) signing a release that authorizes the USGBC to access utility bills directly from utility companies.³⁶²

b. Litigation Resulting from Governmental Legislation and Regulation

There are very few published decisions to date on the issue of green building liability. That said, there have been cases involving individuals, entities, and governments that continue to become more prevalent as more states and cities introduce green building requirements and offer incentives. Many governmental entities have implemented mandatory green building and energy-efficiency standards for new construction and replacement. For example, the City of Albuquerque, New Mexico adopted administrative and technical codes related building and construction, including energy-efficiency standards.

In 2008, a coalition of industry groups and local corporations challenged Albuquerque, New Mexico's new energy-efficiency standards in federal court. In October 2008, the plaintiffs succeeded in obtaining a preliminary injunction against the City of Albuquerque because the court found that federal laws governing energy efficiency preempt municipalities from implementing stricter laws.³⁶³ Later, in January 2012, the court permanently enjoined the city from implementing the municipal efficiency requirements and concluded that all the requirements were preempted by federal law.³⁶⁴

---

360. Press Release, U.S. Green Building Council, Buildings Seeking LEED to Provide Performance Data (June 25, 2009), *available at* http://www.usgbc.org/Docs/News/MPRs%200609.pdf.

361. *See LEED Certification Tools*, U.S. GREEN BUILDING COUNCIL, http://www.usgbc.org/DisplayPage.aspx?cmsPageID=75.

362. Press Release, U.S. Green Building Council, *supra* note 360; *see also* Mireya Navarro, *Some Buildings Not Living Up to Green Label*, N.Y. TIMES, Aug. 30, 2009, at A8, *available at* http://www.nytimes.com/2009/08/31/science/earth/31leed.html.

363. Air Conditioning, Heating & Refrigeration Inst. v. City of Albuquerque, No. 08-633-MV/KBM, 2008 WL 5586316 (D.N.M. Oct. 3, 2008).

364. Air Conditioning, Heating & Refrigeration Inst. v. City of Albuquerque, No. 08-633-MV/KBM (D.N.M. Jan. 25, 2012).

c. Contractual and Other Disputes

Developers, contractors, architects, engineers, and buyers and sellers of real estate are learning to manage the risks and incentives associated with green building initiatives. Many have predicted an onslaught of green building claims based on contract and tort theories. If such claims materialize, entities involved in green building projects may look to professional liability and CGL advertising injury policies to address their emerging risks in the green building area. Insurers should also expect to see more green building claims included among other claims in general construction cases. See Chapter 7, Sections A.3.b and d.

A few cases have been adjudicated in the green building area. One of the first cases involving a dispute over a green building project was initiated in a Maryland state court by the general contractor of 23-unit condominium that was completed in 2006.[365] The developers desired a LEED silver certification for the building and claimed that a project manual attached to the building contract required compliance with a LEED Silver certification. The general contractor filed the complaint for a mechanic's lien in the amount of $54,000. The developer-owners filed a counterclaim for breach of the green building requirements of the contract and negligence, seeking $635,000 in lost tax credits under Maryland's green building program. There is no reported decision in the case because it settled out of court. With the exception of the Maryland and New Mexico cases, there has not been much litigation in this area. Although green building disputes have not been litigated much to date, they are expected to become more common in U.S. jurisdictions.

Green building litigation also may arise between buyers and sellers of green properties. For example, buyers disappointed with their green structures may, in addition to asserting tort and contract claims, bring claims for fraud and unfair competition. Fraud could be in the form of a seller's intentional or negligent misrepresentations in marketing the greenworthiness of a residence. Claims could also take forms similar to those involving greenwashing of products and services. Damages in the context of both green building and greenwashing, however, are a potential obstacle for litigants because they may be difficult to prove and may not be recoverable under state consumer protection statutes.

In *Gidumal v. Site 16/17 Development, LLC*, an owner of a unit of the LEED Gold candidate building, the Riverhouse in New York City, sued the project developer for a variety of construction defects.[366] The plaintiff claimed the unit did not have sufficient heat and also that cold air was entering the building through cracks and holes.

---

365. S. Builders, Inc. v. Shaw Dev., LLC, No. 19 C-07-011405 (Md. Cir. Ct. Somerset County 2007).
366. No. TS-300448-10/NY (N.Y. Cnty.) (transferred to N.Y. Supreme Court).

The plaintiff also alleged fraud and misrepresentation because the unit was not living up to the green standards set forth in the building's offering plan. Thus, the case is a traditional construction defect case, but the plaintiff used green attributes of the building plan to bolster the claim. The fraud and misrepresentation claims, however, may represent a newer trend in the green building area. The plaintiff has appealed an initial decision to the New York Supreme Court, but as of this writing there has been no final decision on appeal.[367]

In August 2011, in *Gifford v. U.S. Green Building Council*, the U.S. District Court for the Southern District of New York dismissed a lawsuit filed by a group of environmental engineering and design professionals against the USGBC for lack of standing.[368] Henry Gifford and his fellow plaintiffs had alleged that the USGBC's false advertising about the energy and money-saving aspects of LEED certification has diverted business from the plaintiffs' businesses to LEED accredited professionals. Plaintiffs also claimed that the false advertising harms consumers. Plaintiffs pleaded claims under the Lanham Act and New York state law. The court dismissed the plaintiffs' Lanham Act claim for lack of standing for several reasons. First, the plaintiffs did not compete with the USGBC in the accreditation of professionals or buildings. Second, the plaintiffs did not allege a reasonable commercial interest that was likely to be damaged by the USGBC's statement that LEED-certified buildings perform on average 25–30 percent better than non-LEED certified buildings. Finally, the chain of causation was speculative. After concluding that there was no standing under the Lanham Act, the court dismissed the remaining state law claims.

These emerging disputes present potential exposures worth analyzing in the CGL advertising liability and professional liability coverage contexts. Insurance companies should expect to see litigation involving green building and greenwashing over issues such as whether materials and products used in construction are indeed "green." Insurers and insureds may enter into disputes with each other regarding whether the failure of a developer or builder to obtain LEED certification, for example, is an event that could be covered under any type of policy. Owners, operators, and tenants in green buildings may sue architects, engineers, and construction contractors if modern green building features (such as green roofs, rainwater collection and storage, renewable energy systems, and use of new or different construction materials) lead to damage of their property. Aside from litigation that may lead to third-party claims, property

---

367. Oral arguments were scheduled for May 7, 2012. Gidumal v. Site 16/17 Development LLC, No. 300448 TSN 2010 (N.Y. App. Div. Apr. 16, 2012).

368. Gifford v. U.S. Green Bldg. Council, No. 10 Civ. 7747(LBS), 2011 WL 4343815 (S.D.N.Y. Aug. 16, 2011).

insurance underwriters also might consider whether their underwriting processes reflect the use of the modern green building features described above.

G. Potential GHG-Related Enforcement Actions

As EPA attempts to use its existing authority under the Clean Air Act to limit GHG emissions, regulatory enforcement actions likely will follow. EPA could also enforce GHG Reporting Rule requirements. Insureds hit with such enforcement actions may try to look for coverage under environmental liability or CGL policies. There are limitations, however, on coverage for such suits. See Chapter 7, Sections A.3.b and c.

CHAPTER 7

Intersection of Climate Change and Insurance

A. Existing and Emerging Climate Change-Related Exposure Arising out of First- and Third-Party Coverage of Insureds ("Potential Claims Liability")

1. Overview of Potential Claims Liability

Insureds faced with climate change-related claims and their insurers must determine whether insureds' existing or historic insurance policies provide coverage for climate change-related claims. It is likely that, in many instances, climate change-related exposures may not have been considered when coverage was written and no premium was collected for climate change-related liability risks. Regardless, existing policy triggers, exclusions, and other policy provisions dictate indemnity and defense obligations under existing policies.

Much of the focus on the impact of climate change has been as a result of episodic climatic events, like hurricanes, involving coverage under first-party policies. As discussed above, many of the early underlying cases have sought relief, not for the individual insured, but for classes of persons requesting remedies such as compelling the government to regulate GHG emissions. Such cases do not implicate liability policies. If the focus shifts from cases seeking additional regulation to claims seeking damages against corporations and professionals based on alleged culpable conduct (tort liability–based allegations), companies and professionals may turn to their liability insurers and seek defense in lawsuits and indemnification for possible resulting settlements and judgments. Where there are third-party impacts from these damages and disruptions, commercial general liability, products liability, environmental liability, professional liability, and directors' and officers' liability insurance may come into play. If the underlying claims are successful and the defense and damages are significant, there also may be implications for excess and umbrella coverage. As the allegations may implicate conduct, knowledge, and damages dating back in time, the potential exists for policyholders to seek coverage pursuant to policies issued years or decades earlier. Resulting defense costs and coverage litigation costs could be high. This section highlights which policy types have been or could be exposed to various types of existing and emerging climate change-related claims.

2. First-Party Coverage

a. Implications for Property Coverage

Episodic climatic events may continue to result in first-party claims under personal and commercial property policies. Property insurers generally have systems in place to monitor how severe weather events—regardless of their potential relation to climate change—are affecting the frequency and severity of first-party property claims. Many insurers and their experts model storm patterns and other natural phenomena without labeling this "climate change" work. This book will not address in detail the risks faced by insurers from such first-party property claims.

b. Implications for Property (Time Element) Coverage

> Climate change-related exposure will be limited in the business interruption scenario to those situations involving physical loss.

Episodic climatic events causing physical loss could potentially expose Property (Time Element) coverages, which are those in which the measurement of loss is tied to a period of time. Examples of the kind of Property (Time Element) coverages that could be affected include business interruption, contingent business interruption, service interruption, and builder's risk. As noted, insurers generally have mechanisms in place to model and track such first-party exposure. Thus, this book focuses on new issues emerging out of the climate change debate, such as regulatory actions pursuant to state environmental policy acts (SEPAs), green building, and use of alternative energy. Although project delays could arise in these areas, climate change-related litigation involving these issues appears unlikely to affect most types of Property (Time Element) coverage because most first-party policies provide coverage only for losses connected to direct physical damage to covered property. Many climate change-related business interruption scenarios, such as delays arising out of a SEPA process, would not result in direct physical loss to an insured's property.[1] On the other hand, climate change-related business interruption scenarios involving physical loss, such as failure

---

1. Some Environmental Policies refer to "Time Element" coverage (for example, Time Element Pollution Event Coverage, which extends coverage to damage caused by pollution that is "abrupt and unexpected"), which is entirely different from, and not to be confused with, "Time Element" coverage in standard first-party property policies.

of green buildings or failure of alternative energy sources (e.g., solar panels damaged in a hail storm), might result in exposure.

(1) Implications for Business Interruption Coverage

Business interruption coverage applies to loss of income during the necessary suspension of operations caused by direct physical loss or damage to insured premises. Business interruption coverage typically includes two fundamental requirements: (1) "direct physical loss or damage" to property at the insured premises, which is caused by (2) a covered cause of loss. The former appears likely to limit coverage for much but not all climate change-related claims. Some scenarios might lead to exposure, such as significant damage to solar panels from an episodic climatic event that leaves a company with no other available power source. In addition, if GHG emissions at or affecting insured premises are deemed to render the property unfit for its intended purpose, coverage might be triggered. In connection with any evaluation of potential business interruption coverage in this context, it bears noting many cause-of-loss forms contain exclusions for "ordinance or law" and "government action," meaning that when reconstruction or repair takes longer to accomplish in order to comply with ordinances or laws or is extended due to other government action, there may not be coverage for resulting lost income or extra expenses incurred (above normal expenses to continue operations).

Thus, although the potential for business interruption exposure related to climate change exists, the requirements for triggering such coverage—including, particularly, the requirement of "direct physical loss or damage"—probably would limit coverage for most climate change-related claims. Insurers nonetheless may want to review and clarify their standard exclusions as well as consider offering new related coverages in this area.

(2) Implications for Contingent Business Interruption and Service Interruption Coverages

Contingent business interruption and service interruption coverages apply to the loss of income or increase in expenses resulting from damage from a covered cause of loss to the premises of another entity on which the insured depends (e.g., suppliers, customers, utilities, and transportation providers). Because such coverage typically is limited to loss caused by "direct physical loss or damage," there is only modest risk of increased exposure to insurers for climate change-related claims. Nevertheless, insurers should review their forms to consider potential exposures.

(3) Implications for Delay in Completion Coverage Under Builders-Risk Policies

Many builders-risk policies provide coverage for delay in completion of construction projects. As with the other time element coverages discussed above, delay in completion coverage is triggered only by direct physical loss or damage to covered property from a covered cause. A collapse of a green roof or failure of some other green building component theoretically could lead to delay resulting in a claim. In addition, if a latent site condition is discovered during construction that has significant carbon footprint impacts and triggers a SEPA or other regulatory action, builders might argue that such a latent condition is damage to the property, therefore triggering coverage. But because much climate change-related litigation, as defined herein, does not appear to involve direct damage to insured builders' property, this coverage is not considered highly relevant to this review. It may have some relevance in the green building context.

3. Third-Party Coverage

a. Implications for D&O Coverage

In recent years, shareholder resolutions related to climate change have been on the rise, and investors have demanded increased disclosure of climate change-related risk. Given the challenges facing plaintiffs in climate change tort litigation, climate change risk disclosure issues may become the focal point of the plaintiffs' bar's efforts rather than traditional tort theories. Regulators are also taking action on climate change disclosure. All these activities may create a new and evolving standard of care for directors and officers in the climate change context. Directors and officers may face actions by regulators and actions by shareholders related to climate change risk. A rise in such claims would have major implications for D&O coverage.

Insureds facing claims related to concealment, misrepresentation, and mismanagement of climate change-related risk may look for reimbursement of defense costs and indemnification pursuant to D&O policies. D&O insurance provides financial protection for an organization's directors and officers in the event they are sued in conjunction with the performance of their duties as they relate to the organization. Many D&O products contain an expanded definition of "claims," including various administrative and regulatory investigations and proceedings. Most D&O policies do not specifically address whether climate change-related claims are covered, although at least two insurers (Zurich and Liberty Mutual) provide some specific coverage in this area. D&O policies may have pollution or pollution-related exclusions.

(1) Actions by Regulators

Some climate change-related claims may arise out of actions by regulators. If the SEC establishes more definitive standards for climate change disclosure in the future, the SEC may pursue companies for noncompliance. The SEC has authority to bring two basic types of enforcement actions: (1) a civil action in federal district court seeking an injunction against further violations or other equitable relief, and (2) administrative proceedings, either in addition to or as an alternative to a federal civil action.[2] As part of either a civil action or administrative proceeding, the SEC may be able to seek penalties and other relief under statutes such as the Remedies Act[3] and the Sarbanes-Oxley Act of 2002.[4] The SEC also has the authority to issue a public report under the Exchange Act, called a section 21(a) report.[5] A section 21(a) report imposes no sanction and is not technically an enforcement action, but the reports may publicly indicate that a corporation or its directors violated securities laws.[6] Generally, section 21(a) reports have only been issued in lieu of an enforcement action and with the agreement of the target, as the report affords no opportunity for a hearing on the merits.[7]

New York attorneys general also may continue to pursue companies for failing to make sufficient climate change-related disclosures under the Martin Act.[8] If climate change regulation gets off the ground in the United States, regulators may pursue companies for mismanaging climate change-related disclosures. Consequently, insurers may see claims for Side A, B, and C coverage[9] stemming from regulatory action against directors and officers and/or the organization itself increase in the coming years. Such claims for coverage will probably include claims for the cost to defend against a regulatory or administrative investigation.

---

2. Brian A. Ochs, Stavroula A. Lambrakopoulos & Rebecca L. Kline Dubill, *Sanctions and Collateral Consequences, in* THE SECURITIES ENFORCEMENT MANUAL: TACTICS AND STRATEGIES 183–84 (ABA 2d ed. 2007).

3. Pub. L. No. 101-429, 104 Stat. 931 (1990); *see also* Ochs et al., *supra* note 2, at 184.

4. Pub. L. No. 107-204, 116 Stat. 745 (2002); *see also* Ochs et al., *supra* note 2, at 184–85.

5. 15 U.S.C. § 78u(a); *see also* Ochs et al., *supra* note 2, at 185.

6. Ochs et al., *supra* note 2, at 185.

7. *Id.*

8. *See* discussion in Chapter 6.E.4 (climate change-related disclosure activity in New York pursuant to the Martin Act).

9. Side A coverage is "[t]he section of coverage under a directors and officers liability insurance policy affording 'direct' coverage of an organization's directors and officers. This portion of the policy provides direct indemnification to the directors and officers for acts for which the corporate organization is not legally required to indemnify the directors and officers." International Risk Management Institute, Inc., Glossary of Insurance and Risk Management Terms, Side A Coverage, *available at* http://www.irmi.com/online/insurance-glossary/terms/s/side-a-coverage.aspx.

Side B coverage, also known as corporate reimbursement coverage, is "[c]overage under a directors and officers (D&O) liability policy covering the corporate organization's obligation to indemnify its directors and officers for claims resulting from their acts in conjunction with the organization." International Risk Management Institute, Inc., Glossary of Insurance and Risk Management Terms, Corporate Reimbursement Coverage, *available at* http://www.irmi.com/online/insurance-glossary/terms/c/corporate-reimbursement-coverage.aspx.

Side C coverage, also known as entity securities coverage, is "[t]he section of a directors and officers (D&O) liability insurance policy covering the corporate entity when claims are made against the entity in conjunction with securities it has issued." International Risk Management Institute, Inc., Glossary of Insurance and Risk Management Terms, Entities Securities Coverage, *available at* http://www.irmi.com/online/insurance-glossary/terms/e/entity-securities-coverage.aspx.

(2) Actions by Shareholders

In addition to increased regulatory action by federal and state authorities, there has been a dramatic increase in shareholder resolutions relating to climate change risk disclosure. Shareholders have filed numerous climate change-related resolutions in recent years. If this trend continues, insurers may eventually receive D&O claims stemming from shareholder derivative suits challenging a corporation's refusal to comply with such resolutions.

In the future, investors also may bring shareholder derivative claims alleging that a corporate director's or board member's failure to consider climate change risk ultimately resulted in a financial loss to the corporation. Directors and officers additionally may face class actions and/or derivative actions by shareholders if a failure to disclose climate change-related risk results in harm to the corporation, such as loss of revenues, loss of market share, falling stock price, reputational damage, and/or missed business opportunities related to climate change issues. Plaintiffs may use evidence from the myriad voluntary and mandatory disclosure programs to make their case. See Chapter 6.E. Directors and officers implicated in such actions will probably look to D&O policies for reimbursement of defense costs and indemnification.

(3) Potentially Relevant Provisions

Arguably, a broad D&O pollution exclusion using "arising out of" language could bar claims arising out of climate change-related harm in a number of jurisdictions. But as noted in Chapter 8.C.2.c, the pollution exclusion is not a silver bullet against exposure for defense expenses, and litigation over its scope is likely to generate substantial expenses. Moreover, some insurers issue D&O policies with modified pollution exclusions that do not apply to mismanagement of pollution but only to more direct acts of pollution. Similarly, the impacts of intentional/criminal acts exclusion are likely to be minimized by (1) disputes over what is the relevant act, (2) severability provisions, and (3) the requirement that intentional/criminal acts be established by judgment or other final adjudication.[10]

The fact that punitive or exemplary damages and most fines or penalties imposed by law are excluded from the definition of loss has some significance in this context, as this may limit indemnity obligations, although not obligations for reimbursement of defense costs for certain types of claims. Another potentially relevant provision is the limitation of liability providing that all loss arising out of the same wrongful act and all interrelated wrongful acts of insured persons shall be deemed one loss subject to one limit of liability provided in the declaration.[11] The bodily injury and property damage

---

10. *See* discussion in Chapter 8.D.9 (intentional/criminal acts exclusion in detail).
11. *See* discussion in Chapter 8.E.2 (batch clauses).

exclusion[12] should preclude coverage for certain kinds of climate change-related tort claims pursuant to D&O policies; such claims are more likely to turn up on CGL or environmental liability policy rosters.[13] If material financial impacts resulted from a climate change tort suit such as *Kivalina*, however, the bodily injury and property damage exclusion likely would not bar shareholder mismanagement claims arising out of the underlying tort suit because many courts interpret the exclusion as barring only direct claims for bodily injury or property damage, thus resulting in potential stacking of D&O and CGL policies.

The intentional/criminal acts exclusion[14] and the definition of loss may have some limited significance to climate change-related claims (coverage litigation and reimbursement of defense costs). There are likely to be disputes over what is the relevant wrongful act, the final and non-appealable adjudication requirement, inability to file a declaratory judgment action as to the wrongful act while the underlying action is pending, and the impact of severability provisions. Potential exposure may be moderated to a limited extent by the definition of loss because punitive, exemplary, and multiple damages as well as most taxes, fines, and penalties are excluded from the definition of loss in the coverage grant.

b. Implications for Commercial General Liability Coverage

Insureds hit with climate change-related tort claims probably will look to commercial general liability (CGL) policies for coverage because CGL insurance generally provides broad coverage for defense and indemnity of claims for bodily injury, personal injury, and property damage (potentially including natural resource damages).[15] In addition, CGL policies generally provide for defense costs outside of limits, meaning the insurer pays both to defend covered claims and to indemnify the insured up to policy limits. Accordingly, CGL coverage is particularly susceptible to insured claims in the climate change context.

The climate change-related litigation that has emerged so far illustrates that a wide range of allegations can potentially trigger CGL coverage, including the duty to defend such claims. For example, allegations of public nuisance, negligence, fraud, and conspiracy claims relating to climate change might trigger CGL coverage. Whether such claims are actually covered by a CGL policy has yet to be fully decided. In an initial win for insurers, the Virginia Supreme Court ruled that Steadfast Insurance Company had no duty

---

12. *See* discussion in Chapter 8.D.1 (the bodily injury and property damage exclusion in detail).
13. There is always the possibility, however, that a claim could have multiple allegations such that it would trigger both D&O and CGL policies.
14. *See* discussion in Chapter 8.D.9.
15. *See* discussion in Chapter 6.A (climate change-related tort litigation).

to defend AES in the *Kivalina* litigation because there was no occurrence within the meaning of the CGL policies.[16] The *AES* litigation[17] illustrates that the following issues are likely to be the subject of ongoing CGL coverage disputes: (1) whether the conduct alleged constitutes an occurrence (pursuant to an objective or subjective standard), (2) the scope of the pollution exclusion, and (3) the application of loss in progress and known loss provisions in the climate change context. See Chapter 8, Sections B.1, C.1.b(2), and D.3 for a detailed discussion of these potential liability drivers.

The use and interpretation of limited pollution coverage endorsements (carving back the pollution exclusion for 30- to 90-day pollution events) also should be considered in light of climate change-related claims arising out of episodic climatic events. In addition, the intentional/criminal acts exclusion may be relevant to fraud and conspiracy cases in the climate change context. See Chapter 8.D.9. Whether damages to natural resources like air, land, and water constitute "property damage" potentially triggering CGL coverage also may be litigated. If injury to such natural resources is considered "property damage" (which typically is defined to include physical damage and/or loss of use), the likelihood for triggering coverage in CERCLA NRD cases related to climate change may increase. See Chapter 6.D, discussing CERCLA NRD claims. Litigation regarding the applicability of all of these exposure drivers in multiple jurisdictions are likely to be costly.

Insureds also may look to CGL coverage for advertising injury liability in the climate change context. For example, builders and architects in the green building industry as well as manufacturers may be subject to greenwashing claims. See Chapter 6.F.3–4.

CGL products liability endorsements also may be implicated by new climate change-related claims. Although no climate change-related products liability claims have been made thus far, it is possible that individuals could bring products-liability claims against an insured, such as failure to warn or design-defect claims related to climate change. Such claims would probably allege that manufacturers failed to warn consumers that products emitted harmful GHGs or that products emitting GHGs are defectively designed. The most probable targets of such claims would be manufacturers of products with high emissions, like automobile manufacturers. A plaintiff could argue that the emission of large amounts of GHGs was avoidable or, in the case of automobile manufacturers, a design defect. An analogous argument could be made against consumer product manufacturers if the product has low energy efficiency and better design or manufacturing options were available.

---

16. AES Corp. v. Steadfast Insurance Co., 725 S.E.2d 532 (Va. 2012). *See* discussion in Chapter 8.B.1 (details of the occurrence issue in *AES*).

17. *Id.*

c. Implications for Environmental Liability Coverage

Coverage for defense and indemnity of claims relating to loss of life, personal injury, and property damage stemming from environmental contamination is typically excluded from CGL policies under the "pollution exclusion." To provide defense and indemnity coverage for claims that fall within CGL pollution exclusions, some insurers offer separate environmental liability insurance. Environmental liability insurance generally covers damages and cleanup costs arising out of "pollution events" at, on, under, or migrating from covered locations and defense costs associated with such events. Whereas a conclusion that GHGs are pollutants for purposes of pollution exclusions could decrease exposure under CGL policies, a determination that GHGs are pollutants might increase exposure under environmental policies. There is also the possibility that a court could read the pollution exclusion narrowly and a coverage grant broadly, thus stacking the policies. Consequently, climate change may present a special risk in the environmental liability insurance field. The fact that most environmental policies are written on a claims-made rather than on an occurrence basis will temper some of this risk because insureds will not be able to look to historic policies. Circumstance reporting of climate change events, however, may become more commonplace as insureds try to preserve their right to demand coverage for climate change events even under claims-made policies. Many environmental policies provide for circumstance reporting.

Insureds may try to seek coverage under environmental liability policies for tort claims arising out of episodic environmental events[18] (see Chapter 6.A.5), other common law tort claims (such as *AEP*, *Kivalina*, and *General Motors*, see Chapter 6.A.2–4), and statutory environmental claims such as natural resource damages under CERCLA (see Chapter 6.D). In addition to tort lawsuits and CERCLA NRD actions, insureds may face statutory liability pursuant to the Clean Air Act. See Chapter 5.B. For example, insureds may attempt to recover costs of compliance through their environmental liability policies. But maintenance and betterment exclusions may prevent exposures for day-to-day operational costs such as those associated with Clean Air Act–related upgrades. See Chapter 8.D.6, discussing that exclusion.

The intentional/criminal acts exclusion could moderate exposures somewhat, because such exclusions in environmental policies generally do not have a final adjudication requirement and some do not have severability provisions. Nevertheless,

---

18. With respect to episodic event claims, plaintiffs may argue that the insured did not act reasonably in preparing for and mitigating the consequences of severe weather allegedly caused by climate change. For example, in *Turner v. Murphy Oil USA, Inc.*, 472 F. Supp. 2d 830 (E.D. La. 2007), Murphy Oil settled a class action lawsuit for more than $330 million where it was alleged that Murphy Oil acted unreasonably with regard to tank safety measures and that Murphy Oil knew, or should have known, that the tank would rupture in a hurricane-related flood.

coverage battles over the applicability of this exclusion could be costly. See Chapter 8.D.9. The fines, penalties, and punitive damages exclusion could provide some moderation of exposure for statutory fines and penalties related to climate change, if any, and claims seeking punitive damages. See Chapter 8.D.7–8. Limits of liability provisions for multiple claims arising out of the same pollution event, multiple insureds or claimants, multiple coverages, multiple policy periods, and sublimits probably will not eliminate exposure but could help to manage exposure within certain policy limits. See Chapter 8.E.

d. Implications for Professional Liability/E&O Coverage

Professional liability insurance covers errors and omissions of businesses and their professionals: engineers, architects, consultants, and accountants. Moreover, it is distinguishable from CGL insurance in that it covers indemnification and defense of claims by third parties against an insured for damages, including economic damages, arising out of errors and omissions in work performed for a third party in breach of a service or work agreement. Insurers should be alert to the increasing risk of claims arising out of a professional's errors in performing climate change-related work. Already, there has been at least one lawsuit related to failures of a contractor to deliver a LEED-certified green building. See Chapter 6.F.4. Climate change-related claims could arise out of a wide variety of professional activities. As EPA, states, and nonprofits increasingly demand disclosure of GHG emissions, professionals may be called on to account for or calculate GHG emissions for corporate clients for reports to various authorities. For example, companies may rely on these professionals to help prepare reports pursuant to the EPA GHG Reporting Rule (see Chapter 5.B.4) or in connection with preparation of environmental impact statements related to obtaining permits at the state and local level (see Chapter 6.F.1). If professionals make mistakes in the course of doing such work, their customers may file claims against them. Professionals also could be subject to liability if there are problems arising from misrepresentations or mistakes made in connection with green building, installation of emissions reduction solutions, and possibly handling of carbon credits. See Chapter 6.F.2 and 4. For example, if a professional fails to take the promised steps to mitigate GHGs in a "green building" project, causing a loss of LEED certification or loss of tax benefits, the client may have a claim against the professional.

Professionals will probably seek coverage for these kinds of climate change-related claims. Whether such claims will be covered will depend on the coverage and exclusions contained within the policy. A pollution exclusion may or may not apply in this circumstance, even if the policy has a pollution exclusion. Professional liability policies typically contain intentional/criminal act exclusions, but the impact of such exclusions could be tempered by severability provisions. The scope and impact of (1) fines, penalties, and taxes exclusions; (2) punitive, exemplary, and multiple damages exclusions; and possibly (3) products liability exclusions may be relevant in the climate change context. In particular, a fines or penalties exclusion could temper the exposure from claims involving professionals who were penalized for violating EPA's GHG Reporting Rule. Finally, limits of liability for multiple claims relating to the same negligent act, error, or omission may reduce but not eliminate exposure.

e. Implications for Contractor's Protective Professional Indemnity and Liability Insurance and Owner's Protective Professional Indemnity Insurance

Some insurers offer owners and contractors professional indemnity insurance to supplement their primary professional liability policies. Owners protective professional indemnity insurance (OPPI) provides first-party indemnity coverage for construction-project owners that pays based on an established third-party liability loss that arises out of the owner's subcontracted design services. Such coverage generally is excess over any available primary professional liability insurance (of a noninsured) or the insured's own self-insured retention (SIR). OPPI policies also may provide third-party liability defense coverage should the owner incur a loss from a third party attributable to the negligence of its design professionals.

Contractors protective professional indemnity and liability insurance (CPPI) is similar to the coverage offered by OPPI. CPPI protects contractors for losses arising out of the negligent performance of professional services for which the contractor is responsible (such as compliance with green building, SEQA, EIS, and/or Environmental Impact Review requirements), whether directly or indirectly, incidentally, or on a contingent or vicarious basis. A CPPI policy generally provides two coverages: (1) third-party professional liability coverage and (2) first-party professional indemnity against an uninsured or underinsured design professional (a noninsured under the policy) under contract with the insured. In addition, contractor's pollution coverage that includes disposal sites, mold liability, and transportation coverage is often available by endorsement. See also Chapter A.3.c discussing environmental liability policies.

Because OPPI and CPPI provide first-party coverage for mistakes made by subcontractors, OPPI and CPPI coverages may have exposure to climate change-related risks similar to the exposure to third-party professional liability/E&O coverage. Professional services–related climate change claims could arise out of a wide variety of professional activities, such as preparation of GHG emissions reports, SEQA activities, preparation of environmental impact statements, and green-building activities, including LEED certification.[19] OPPI and CPPI coverages will indemnify the insured for any damages paid by the owner insured or contractor insured to the third party or for the loss suffered by the owner insured or contractor insured caused by the design professional's errors or omissions that are in excess of the "losses" paid to the insured by the design professional's own professional liability insurer.

Whether an insurer will be required to indemnify an owner or contractor under an OPPI or CPPI policy for losses caused by a design professional's own errors or omissions will depend in part on the exclusions contained within the OPPI or CPPI policy. Significantly, like third-party professional liability policies, OPPI and CPPI policies may or may not contain pollution, GHG, or climate change exclusions. These policies may contain other exclusions typically found in professional liability policies, such as exclusions for fines and penalties, multiple damages, and product liability, which may have some relevance to reducing exposure for climate change-related claims. In addition, OPPI policies often exclude from coverage dishonest, fraudulent, criminal, intentional, or malicious acts, errors or omissions committed by, between, or at the direction of the design professional or the named insured.

4. Potential for Bad Faith Claims

a. Bad Faith Claims Based on Inconsistent Position on Climate Change-Related Issues

If climate change-related litigation increases and insurers begin to take coverage positions on the sundry climate change-related claims discussed above, insurers should expect extra-contractual bad faith claims to follow. Representations and decisions made by insurers before climate change-related case law and legislation is fully developed could give rise to bad faith claims if caution is not exercised. Conflicting statements from corporate, marketing, claims, and underwriting units on climate change issues could lead to allegations of intentional misrepresentation and claims discrimination. Communication and coordination among the various corporate units on climate change issues may help avoid any inadvertent inconsistencies or misrepresentations.

b. Bad Faith Claims Based on Denial of Coverage

Even when coverage decisions are consistent, insureds may still pursue claims of bad faith, as the payouts on bad faith claims are often substantial and can exceed policy limits. Generally, insureds will prevail on such claims only when the insurer is aware at some point in the claims handling process that the claim is probably covered. Many

---

19. For a detailed discussion of how third-party professional liability policies, including CPPI, may be exposed to climate change-related risk, see Chapter 6.F.

courts apply a "gross negligence" or "reckless disregard" standard.[20] Under such a standard, it is questionable whether a court could find that a claim alleging coverage for losses arising out of climate change was not "fairly debatable,"[21] especially since the available case law does not directly address such claims. As the jurisprudence develops, however, it may present claimants with an opening to allege that insurers knowingly and in bad faith denied coverage for climate change-related claims.

B. Emerging Climate Change-Related Exposure: Potential Enterprise Liability

1. Potential Liability Based on Corporate Disclosures

If an insurance company is listed on the U.S. Stock Exchange, it is subject to SEC reporting requirements. Accordingly, any regulations or further guidance that the SEC may ultimately issue regarding climate change risk–related disclosures could affect such an insurer. See Chapter 6.E.3.b, discussing SEC guidance on climate change disclosures.

Certain states have decided to administer the NAIC Climate Risk Disclosure Survey to insurers on either a mandatory or voluntary basis. See Chapter 5.H.1, discussing the NAIC survey in detail. Some states are making insurers' responses to the NAIC disclosure survey publicly available. In the past, insurers were able to respond to the host of voluntary surveys without making their full submissions public.

Given that investors, regulators, and the plaintiffs' bar are likely to review insurers' responses to any climate change-related surveys, insurers should carefully prepare such responses and consider issues such as the protection of confidential or proprietary information. A third party may treat a disclosure as an admission and try to use it against the company and its customers. Similarly, any omission or misstatement also could subject the company to unwanted attention of state regulators or litigation with the company's insureds.

....

> Statements made in voluntary disclosures should be consistent because such disclosures could be used against an insurer, which could result in corporate liability.

....

Inconsistent or incomplete statements made by or attributable to an insurer regarding climate change issues could result in potential legal liability. An insurer also

---

20. *See* Jessen v. Nat'l Excess Ins. Co., 776 P.2d 1244 (N.M. 1989); Aetna Cas. & Sur. Co. v. Day, 487 So. 2d 830 (Miss. 1986).
21. *See* Clark-Peterson Co. v. Indep. Ins. Assocs., 514 N.W.2d 912 (Iowa 1994); Erie Ins. Co. v. Hickman, 622 N.E.2d 515, 520 (Ind. 1993); Palmer by Diacon v. Farmers Ins. Exch., 861 P.2d 895 (Mont. 1993); Rawlings v. Apodaca, 726 P.2d 565 (Ariz. 1986).

could face exposure as a result of its participation in ClimateWise or similar climate change-related organizations. Thus, insurers should carefully consider what organizations they participate in and try to make sure that all statements are accurately attributed to their sources, especially if the insurer does not agree with them.

2. Potential Liability Related to Provision of Loss Control Consulting Services

As insurers continue to offer and expand their enterprise risk-management services in the climate change area, insurers should be mindful that holding themselves out as experts in climate change risk management may potentially expose insurers to claims related to errors and omissions of their risk-management consultants. As noted in Chapters 6.E.5 and 7.A.3.d, professionals hired to provide climate change and green building–related services may be subject to liability for problems arising out of misrepresentations or mistakes made in connection with preparation of carbon footprints, GHG mitigation plans, green building designs, and/or efforts to obtain carbon credits. For example, if a professional failed to take the promised steps to mitigate GHGs in green building, causing a loss in LEED certification or loss of valuable credits/offsets, the customer may make a claim against the professional for, among other things, breach of contract or negligence. Insurers may have potential claims liability related to these emerging claims, and this could also be an enterprise liability issue if an insurer's loss-control consulting arm offers these kinds of services.

Risk management consultants at insurance companies should consider obtaining coverage applicable to the delivery of loss-control services and the enterprise at large. The implications of making claims related to that coverage should be evaluated to ensure that the insurer is not taking any inconsistent positions on interpretations of pollution exclusions or other policy provisions within the enterprise for its own coverage or that of its insureds. Risk-management consultants also should be well schooled in relevant and evolving standards of care set by organizations such as the International Organization for Standardization (ISO) as well as regarding regulations in burgeoning areas such as estimating GHG emissions and green building. Ongoing training is necessary in an emerging risk area. Loss-control consulting contract and standard report language also should be reevaluated and updated in light of these emerging risks.

C. New Applications of Existing Coverages and New Products

Development, production, and use of renewable energy is growing as economies seek alternatives to the combustion of fossil fuels that emit GHGs. See Chapter 5.E (discussing renewable energy trends). Traditional insurance products, such as CGL and property policies, may assist various parties involved in renewable energy with risk management. Insurers also are developing new products specific to the renewable energy industry to satisfy a growing need for tailored coverage and to address coverage gaps.

1. Traditional Insurance Products for Renewable Energy Projects

Depending on the nature of a renewable energy project, traditional property and casualty policies may be available to help manage the risk of property damage to the energy asset or other losses, such as personal injury or business interruption.[22] These insurance products generally are available to the whole "lifecycle" of renewable energy production, from solar panel manufacturers to renewable energy producers.[23] Such insurance products provide broader coverage or are more widely available, however, for commercial renewable technologies located in comparatively benign environments such as onshore wind and small-scale hydro projects.[24] Traditional insurance coverages generally are less widely available for offshore wind projects and wave and tidal power technologies, although insurers are beginning to close the coverage gap in these areas.[25] Coverage may not be available for several reasons, including insurers' general concerns about underwriting new or prototypical technology; the inherent technical perils associated with some kinds of renewable energy projects; the inherent political risk in the underlying business model; and faults in design, material, and workmanship.[26]

a. Property Insurance

Traditional property insurance policies may cover damage to renewable energy project assets from extreme weather events, such as hurricanes, hail, floods, and fires.[27] In addition, these policies may compensate the insured against theft of system components, which is a particular concern for the solar industry given the risk of panels

---

22. ELLIOT JAMISON & DAVID SCHLOSBERG, CalCEF, INSURING INNOVATION: REDUCING THE COST OF PERFORMANCE RISK FOR PROJECTS EMPLOYING EMERGING TECHNOLOGY, 12 (Oct. 2011).
23. *Products and Services*, SOLARINSURE, http://www.solarinsure.com/products-and-services.
24. UNITED NATIONS ENVIRONMENT PROGRAMME, SURVEY OF INSURANCE AVAILABILITY FOR RENEWABLE ENERGY PROJECTS 7–8 (March 2006); *see also* JERRY DOUGLAS & ELDRED CLARK, RENEWABLE ENERGY: THE INSURANCE CHALLENGE, AN ACE EUROPEAN GROUP WHITE PAPER 5–7 (2011) (discussing renewable technologies and associated risks).
25. *Cf.* UNITED NATIONS ENVIRONMENT PROGRAMME, SURVEY OF INSURANCE AVAILABILITY FOR RENEWABLE ENERGY PROJECTS 9 (March 2006); *GCube Insurance Services, Inc. Expands BioFuels Team*, MARKET WIRE, May 23, 2011, at 1.
26. UNITED NATIONS ENVIRONMENT PROGRAMME, SURVEY OF INSURANCE AVAILABILITY FOR RENEWABLE ENERGY PROJECTS 9–10 (March 2006); *see also* BETHANY SPEER, MICHAEL MENDELSOHN & KARLYNN CORY, NAT'L RENEWABLE ENERGY LAB., INSURING SOLAR PHOTOVOLTAICS: CHALLENGES AND POSSIBLE SOLUTIONS 23 (Feb. 2010) (noting continued challenges in obtaining affordable insurance for the renewable energy industry, including brokers' and underwriters' unfamiliarity with renewable energy technologies and associated risks).
27. JAMISON & SCHLOSBERG, *supra* note 22, at 12.

being stolen before they are fully affixed during construction.[28] Insurers in this space may offer property insurance for a wide range of renewable energy projects, including wind, solar, biomass, geothermal, and hydropower.[29] Coverage designed to protect a building and renewable energy project during construction typically is referred to as a "builder's risk" policy.[30]

These policies also may provide protection against business interruption losses associated with extreme weather events, such as when the renewable energy asset is out of service due to a hurricane.[31] This kind of coverage often is required to protect the cash flow of the project, including by providing compensation for lost sales and production-based incentives.[32] It can be particularly important when projects are financed under third-party ownership structures.[33] Business interruption coverage, however, may only be available when the business interruption is associated with physical damage.

b. Boiler and Machinery Insurance

Boiler and machinery insurance, which is also known as equipment breakdown insurance, covers losses due to equipment failure from accidental causes and typically does not cover equipment defects that would be addressed by a product warranty.[34] It potentially can address a critical coverage gap by covering direct physical losses arising from a mechanical breakdown, which in the case of solar photovoltaic systems might include occurrences such as ruptures or bursting, electric arcing, or explosions.[35] Boiler and machinery insurance also may provide compensation for business interruption associated with the breakdown of energy assets.[36] Boiler and machinery insurance, however, commonly excludes prototypical technologies or equipment, along with equipment that lacks a sufficient number of in-service operating hours.[37]

---

28. SPEER ET AL., *supra* note 26, at 13.
29. ZURICH FIN. SERVS. GRP., THE CLIMATE RISK CHALLENGE: THE ROLE OF INSURANCE IN PRICING CLIMATE-RELATED RISKS 7 (2009); *see also* JERRY DOUGLAS & ELDRED CLARK, RENEWABLE ENERGY: THE INSURANCE CHALLENGE, AN ACE EUROPEAN GROUP WHITE PAPER (2011) (discussing challenges of insuring various types of renewable energy projects).
30. KATHLEEN ZIPP, *The Excessive and Essential in the World of Risk Management: Project Construction*, WIND POWER ENG'G & DEV., Apr. 11, 2012, *available at* http://www.windpowerengineering.com/construction/the-excessive-and-essential-in-the-world-of-risk-management-project-construction; Robert L. Sobel, *Safeguarding Solar Projects: New Insurance Offerings Require Close Examination*, SOLAR INDUSTRY, January 2011, at 1.
31. SPEER ET AL., *supra* note 26, at 14–15.
32. *Id.*
33. *Id.* at 18.
34. JAMISON & SCHLOSBERG, *supra* note 22, at 12.
35. Sobel, *supra* note 30, at 2.
36. JAMISON & SCHLOSBERG, *supra* note 22, at 12.
37. *Id.*

c. Commercial General Liability Insurance

Commercial general liability policies may protect owners and participants in renewable energy projects from claims of liability for property damage or bodily injury allegedly caused by the insured and arising out of involvement with the project.[38] These kinds of policies typically require an insurer to provide a defense or reimburse the insured's legal costs stemming from a lawsuit.[39] Exclusions in standard policies, however, might limit their efficacy for certain types of claims. For example, in a case where water damage resulting from leaks in improperly designed, manufactured, or installed solar panels leads to mold or fungi claims, the typical commercial general liability policy will exclude coverage.[40]

2. Insurance Policies Tailored for the Renewable Energy Industry

In addition to traditional insurance products, insurance companies have begun offering specialized coverages tailored to the renewable energy industry. Indeed, the insurance market for businesses in the industry has increasingly become more competitive, and insurers increasingly are responding by setting up specialized renewable energy units and developing in-house teams of experts to conduct the necessary underwriting due diligence.[41] One of the most commonly available policies is product liability insurance for renewable products such as wind turbines, photovoltaic cells, and inverters, among other products.[42] Product-liability insurance protects manufacturers from warranty claims regarding defects that cause bodily injury, result in business interruption, and/or associated contractual disputes.[43] While this type of coverage generally is targeted for manufacturers, such policies also may benefit other stakeholders, including developers and investors.[44]

Solar panel warranty insurance is another risk transfer solution that is becoming more widely available. One type of product provides solar panel replacement coverage that is designed to wrap around or replace the manufacturer's warranty.[45] A limited number of insurers recently have begun to provide a different type of

---

38. *Id.*
39. *Id.*
40. Sobel, *supra* note 30, at 2.
41. Sheena Harrison, *Insurers Adapt to Cover Clean Energy Business: Constant Changes Create Challenges in Creating Policies*, Bus. Ins., Feb. 27, 2012, at 2.
42. *See, e.g., Product Liability Insurance for Solar Manufacturers*, SolarInsure, http://www.solarinsure.com/product-liability-insurance-for-solar-manufacturers; Press Release, ET Solar, ET Solar Group Announces Worldwide Product Liability Insurance Coverage (July 7, 2009), www.etsolar.com/mediaCenter/Detail_40.html.
43. *Product Liability Insurance for Solar Manufacturers*, SolarInsure, http://www.solarinsure.com/product-liability-insurance-for-solar-manufacturers.
44. Jamison & Schlosberg, *supra* note 22, at 13.
45. *See, e.g., Solar Panel Warranty Insurance*, SolarInsure, http://www.solarinsure.com/solar-panel-replacement-insurance.

warranty product, known as solar photovoltaic output insurance, which transfers to the insurer the risk of a panel's failure to produce a guaranteed energy output.[46] A principal coverage concern for the latter type of policy is determining whether an event constitutes a loss that will trigger coverage. Among other things, both insurers and insureds must understand whether a covered loss would be based on the annual energy output or a shorter time period, as well as whether coverage is for the entire project or a particular component part.[47]

Another type of specialized insurance policy provides coverage for system performance risks associated with the installation of renewable energy technologies.[48] Similar to solar photovoltaic output insurance, these policies help to protect renewable-energy project owners from specific technology performance risks, such as situations in which the technology does not perform as expected or fails to produce either the projected quantity or the projected quality of output. These policies also may help to address the economic impact to a project if energy or power utilization significantly exceeds the level specified.[49] Policies marketed to renewable energy projects and other businesses in the renewable energy industry might include (or offer the option to obtain) coverage for debris removal and recycling expenses and "green" upgrades for damaged equipment or property. Insurers also are beginning to offer multiline solutions that combine various types of coverages of interest to renewable-energy project developers.

---

46. Trevor D. Stiles & Benjamin P. Sykes, *Insuring Renewable Energy Projects*, LAW360, May 6, 2011, at 2.
47. *Id.* at 3.
48. JAMISON & SCHLOSBERG, *supra* note 22, at 14.
49. Interview with Mike Smith, *Risk Management in the PPA Marketplace*, ALTENERGY eMAGAZINE, *available at* http://www.altenergymag.com/emagazine/2010/08/altenergymag-interview—risk-management-in-the-ppa-marketplace/1556.

CHAPTER 8
Potential Claims Liability: Potential Exposure Drivers

A number of policy definitions, exclusions, coverage grants, limitations, and general conditions are potential exposure drivers in the climate change context. The potential impact of these potential exposure drivers is discussed in detail below.

Commercial General Liability	• "Occurrence" Coverage
	• "Claims Made" Coverage
	• Pollution Exclusion
	• Other Exclusions:
	  • Expected or Intended
	  • Known Loss, Known Injury or Damage, Loss in Progress Clauses
	  • Material Published with Knowledge of Falsity (advertising injury)
	  • Breach of Contract (advertising injury)
	  • Quality or Performance of Goods (advertising injury)
	• Limits of Liability:
	  • Batch Clauses
	  • Defense Costs—Outside Limits
	• Definitions:
	  • "Pollutant"
	  • "Occurrence"

Environmental Liability	• "Claims Made" Coverage
	• Potential Claims Reporting
	• Coverage Grants—Pollution
	• Exclusions:
	  • Pre-Existing Conditions or Known Pollution Events
	  • Intentional/Criminal Acts
	  • Fines, Penalties, Punitive Damages
	  • Naturally Occurring Substances
	  • Maintenance/Betterment
	  • Injunctive Relief
	• Limits of Liability:
	  • Multiple Insureds, Claims, Coverages, Policy Periods
	  • Sub-limit
	• Definitions:
	  • "Claim"
	  • "Cleanup Costs"
	  • "Natural Resource Damages"
	  • "Pollutant"
	  • "Pollution Event"

Directors and Officers Liability	• "Claims Made" Coverage
	• Circumstance Reporting
	• Coverage Grants—specific climate change-related coverage
	• Pollution Exclusion
	• Other Exclusions:
	  • Intentional/Criminal Acts
	  • Related Wrongful Acts
	  • Bodily Injury & Property Damage
	• Definitions:
	  • "Loss" (whether it includes fines, penalties, punitive damages)
	  • "Pollutant"
	• Limits of Liability and Retentions/Deductibles:
	  • Related Claims

Professional Liability	• "Occurrence" Coverage
	• "Claims Made" Coverage
	• Economic Damage Coverage
	• Circumstance Reporting
	• Exclusions:
	  • Pollution Exclusion
	  • Intentional Acts
	  • Fines, Penalties, Punitive Damages
	  • Products Liability
	• Limits of Liability:
	  • Multiple Claims

A. Claims-Made and Reported

A claims-made and reported trigger is most common in professional liability, D&O, environmental liability, and umbrella/excess policies. There are also some claims-made CGL policies. Because all of these types of policies are susceptible to climate change-related claims, it is important to understand the implications of a claims-made trigger.

Under a claims-made policy, two conditions generally must be present to trigger coverage: (1) the claim must be made during the policy period or the extended reporting period and (2) the incident that led to the claim must have taken place, or occurred, on or after the policy's retroactive date. Some policies also allow a particularized circumstance that does not rise to the level of a claim to be reported during the policy period. If a particularized circumstance is reported within the policy period, a later claim based on that circumstance may be covered. Some claims-made policies contain retroactive dates that extend back to the date of the insured's initial operations while others provide the inception date of the policy as the retroactive date. In addition, claims-made policies sometimes do not contain any retroactive date limitation.

Accordingly, while exposure might not be more likely under historic occurrence-based policies, a claims-made trigger will not eliminate exposure to climate change-related claims due to availability of circumstance reporting, absence of a retroactive date, or inclusion of a retroactive date that extends back to the beginning of operations. Below is a discussion of how each of the claims-made trigger conditions may be relevant when insureds seek coverage for climate change-related claims.

1. The Claim Must Be Made During the Policy Period or Extended Reporting Period

While there are some variances, under most claims-made policies a claim is "made" when it is first presented to the insurer. Although this may sound simple, the challenge occurs in determining what constitutes a claim under the policy. A verbal allegation may satisfy one policy, whereas another policy may require a written demand. In addition, most claims-made policies provide coverage only if the claim is "first made" during the policy period. This precludes coverage for claims that should be the responsibility of the previous insurer and prevents the insured from obtaining coverage under a renewal policy that may have more favorable terms. Also, under the typical claims-made and reported policy, a claim is not deemed "made" until the insured provides notice of the claim to the insurer. Some claims-made policies, however, only require the claim to be reported "as soon as practicable." In the climate change context, it will be important to determine exactly when the insured was first made aware of the claim, something that could be particularly difficult in situations where activist groups make repeated demands to an insured to reduce or stop emitting GHGs. For instance, in the D&O context, Ceres has been very active in demanding that corporations provide full climate change disclosure in their SEC filings.

In addition to covering claims made during the policy period, some policies will cover a future claim if the circumstances of the claim were reported during the policy period. Professional liability, environmental liability, and D&O policies contain circumstance reporting provisions providing that as long as a particular circumstance leading to a future claim is reported during the policy period, the claim will be deemed to have been made when the circumstance was reported as opposed to later when the actual claim arises. Given the circumstance reporting provisions in these policies, insureds could start to report GHG-related "circumstances" in the coming years in order to preserve coverage. If insureds are concerned this may lead to rising premiums in future years, however, they may avoid doing so until absolutely necessary. Other insureds may be monitoring the first round of cases to determine if such reporting would be worthwhile.

2. The Retroactive Date Requirement

The retroactive date requirement in some claims-made policies (principally in CGL and D&O policies) mandates that the "incident" that leads to the claim have taken place, or occurred, on or after the policy's retroactive date. In climate change-related cases, the "incident" may be the continuous emission of GHGs. Determining when the damage-causing "incident" occurred will be critical to determine whether the policy is triggered. If the continuous emission of GHGs is considered an "incident," insureds may argue that the "incident" occurred both before and after the retroactive date, triggering coverage under the claims-made policy and possibly multiple prior policies if the latter are occurrence-triggered policies. Determining which policies are triggered could lead to potentially expensive coverage litigation. Retroactive dates, however, may provide underwriters with tools for managing potential exposures and premiums.

B. Occurrence-Based

Whether a policy is occurrence-based is an important exposure driver in the climate change context. Many CGL policies are occurrence-based. Under an "occurrence" policy, coverage is triggered if there is an "occurrence" during the policy period. The claim can be made long after the policy period. Thus, insureds can reach back to historic policies to look for coverage. "Occurrence" is typically defined to mean "an accident, including continuous or repeated exposure to substantially the same general harmful conditions."[1]

What constitutes an "occurrence" to trigger coverage in the climate change area will be hotly contested. Already, whether the emission of GHGs can be considered an "occurrence" was litigated in Virginia in *AES Corp. v. Steadfast Insurance Co.*,[2] and an Indiana court has considered the occurrence issue in an emissions-related case, *Cinergy Corp. v. Associated Electric & Gas Insurance Services, Ltd*. This section discusses those two cases, the potential effect of various trigger doctrines, and long-tail claims.

> Whether a policy is occurrence triggered is a significant issue in the climate change context.

1. *AES Corp. v. Steadfast Insurance Co.*

In the *AES* litigation, Steadfast sought a declaratory judgment that it had no duty to defend or indemnify AES in relation to the underlying *Kivalina* lawsuit.[3] Steadfast

---

1. ISO Form CG 00 01 12 07.
2. AES Corp. v. Steadfast Ins. Co., 725 S.E.2d 532 (Va. 2012).
3. Native Vill. of Kivalina v. ExxonMobil Corp., No. 09-17490, ___ F.3d ___, 2012 WL 4215921 (9th Cir. Sept. 21, 2012), *petition for rehearing* filed Oct. 4, 2012. *See* Chapter 6.A.3.

had denied coverage based on three grounds, including that the *Kivalina* complaint did not allege "property damage" caused by an "occurrence" as those terms were defined in its policies.[4] The policies defined "occurrence" as "an accident, including continuous, repeated exposure to substantially the same general harmful condition." Steadfast argued that the *Kivalina* complaint does not allege an accident because it alleges intentional actions and foreseeable consequences.[5]

AES argued that, because the complaint alleged that AES "[i]ntentionally or negligently" created a nuisance, the damage alleged by plaintiffs in *Kivalina* constitutes an "accident" and thus an "occurrence." AES further argued that because the *Kivalina* complaint alleges that AES "knew or should have known" that generation of electricity would result in the environmental harm suffered by Kivalina, the complaint alleges, at least in the alternative, that the consequences of AES's intentional CO_2 emissions were unintended. The trial court granted summary judgment for Steadfast, holding that the insurer had no duty to defend AES because the *Kivalina* complaint did not allege an occurrence within the meaning of the CGL policies issued to AES by Steadfast. The Virginia Supreme Court affirmed the decision that the *Kivalina* case alleged no occurrence. Then, in a rare move, on January 17, 2012, the court granted a petition for rehearing and vacated the decision.[6]

In the petition for rehearing, AES argued that the authorities cited by the court in support of its initial decision did not support its ruling. In evaluating whether there is an occurrence, AES urged the court to differentiate between allegations that a defendant should have known that harm was "reasonably foreseeable" and allegations that a defendant should have known that there was a "substantial probability" that harm would occur.[7] AES argued that the cited authorities only excuse the duty to defend when the complaint alleges that the defendant "should have known to a substantial probability" that its conduct would cause the harm alleged.[8] The Virginia Trial Lawyers Association filed a motion seeking permission to file an amicus brief in support of rehearing because, in its view, "the court's opinion as currently written likely wipes out liability coverage for most negligence-based claims."[9]

On rehearing, the Supreme Court of Virginia again affirmed that Steadfast had no duty to defend AES in connection with Kivalina's claims. In a second opinion by Justice Bernard Goodwyn, the court again held that the underlying complaint against AES did not allege an occurrence. The court held that coverage is precluded under

---

4. For a discussion of the two other grounds for denying coverage, see Chapter 8, Sections C.1.b.2 (meaning of "pollutant" and the pollution exclusion) and D.3 (loss in progress clauses).
5. Brief of Appellee at 19, AES Corp. v. Steadfast Ins. Co., No. 100764 (Va. Oct. 8, 2010).
6. Order, *AES Corp.*, No. 100764 (Va. Jan. 17, 2012) (granting petition for rehearing).
7. Petition for Rehearing at 4, *AES Corp.*, No. 100764 (Va. Oct. 17, 2011).
8. *Id.*
9. Brief Amicus Curiae of Virginia Trial Lawyers Ass'n, *AES Corp.*, No. 100764 (Va. Oct. 17, 2011).

a CGL policy where the underlying complaint alleges that the result of the insured's intentional act "was a natural or probable consequence of the intentional act."[10] The court explained that even if an insured is negligent and did not intend to cause the alleged damage, if an insured knew or should have known that certain results were the "natural or probable consequences of intentional acts or omissions," there is no accident or occurrence under a CGL policy.[11]

The decision did not address AES's "substantial probability" argument. Instead, the court articulated a different standard. The court stated: "For coverage to be precluded under a CGL policy because there was no occurrence, it must be alleged that the result of an insured's intentional act was more than a possibility; it must be alleged that the insured subjectively intended or anticipated the result of its intentional act or that objectively, the result was a natural or probable consequence of the intentional act."[12] In *AES*, the court found that there was no occurrence because Kivalina alleged that the natural and probable consequence of AES's emissions was global warming and the harm it suffered as a result along its coastline.

The Virginia Supreme Court decision in *AES* is an important step toward resolving a key issue in climate change-related coverage litigation—whether there is an occurrence for purposes of a CGL policy. In future climate change-related coverage litigation, both insurers and policyholders should be mindful of whether the relevant jurisdiction applies an eight corners approach to determining coverage and whether the court applies an objective or subjective standard to interpreting the term "occurrence." The *AES* decision resulted in part from the rule that only the allegations of the complaint and policies should be considered in determining whether there is a duty to defend (i.e., the eight-corners rule) and Virginia's application of an objective standard to evaluating whether there is an occurrence. The *AES* case accepted the allegations of the underlying *Kivalina* complaint regarding climate change as true, even though some still dispute whether global warming is a scientific certainty. Moreover, the court focused on the complaint's allegations of what AES knew or should have known about the potential impact of its emissions, not on what AES subjectively believed would be the consequence of its emissions. In future climate change-related coverage disputes, whether the jurisdiction applies the eight-corners rule and an objective or subjective standard for interpreting whether there is an occurrence may be relevant.

---

10. AES Corp. v. Steadfast Ins. Co., 725 S.E.2d 532, 536 (Va. 2012).
11. *Id.* at 538.
12. *Id.* at 536.

2. *Cinergy Corp. v. Associated Electric & Gas Insurance Services, Ltd.:* Damages Not Caused by An Occurrence

Cinergy Corp. v. Associated Electric & Gas Insurance Services, Ltd. provides a good illustration of the type of coverage litigation that could arise in the climate change arena.[13] In *Cinergy*, the insured, a power company, sought defense costs from its excess insurer to defend against a federal suit by EPA seeking to enjoin the insured from future violations of the Clean Air Act and to require the installation of modern pollution control technology to prevent future unlawful emissions. The case raised the question of whether there was an occurrence. The principal disagreement between the insured and the insurer was whether the costs of installing new equipment fell within the policies' coverage for damages because of or resulting in bodily injury or property damage with respect to any accident.[14]

The court held that prophylactic measures meant to prevent future emissions are not damages caused by an "accident" within the meaning of the term "occurrence," but instead resulted from the prevention of such an occurrence. Consequently, because the underlying lawsuit did not allege an occurrence for which damages were sought, there was no coverage under the policy.

Cinergy appealed the decision to the Court of Appeals of Indiana, this time claiming that the costs associated with Cinergy's surrender of emissions allowances, court-imposed penalties, and payment of attorney fees should be covered under its CGL policy. The court of appeals ruled in favor of the insurer. With regard to the costs of surrendering emissions allowances, the court held that "[c]ourt-ordered remedies imposed to prevent future emissions 'are not caused by the happening of an accident, event, or exposure to conditions, but rather result from the prevention of such an occurrence.'"[15] Cinergy claimed that the penalties and attorney fees were compensatory damages and thus covered under its CGL policy. The court again held that because there was no occurrence under the terms of the policy, the penalties and attorney fees are not covered under the insurers' policies.[16]

Cinergy also asked the court to postpone its determination of Cinergy's right to indemnification under its policies until resolution of the underlying litigation. The court of appeals refused to do so and upheld the trial court's determination that the insurers had no duty to indemnify Cinergy, again reasoning that no occurrence or potential occurrence triggering coverage under the policies took place.[17]

---

13. Cinergy Corp. v. Associated Elec. & Gas Ins. Servs., Ltd., 865 N.E.2d 571 (Ind. 2007).
14. *Id.* at 579.
15. Cinergy Corp. v. St. Paul Surplus Lines Ins. Co., 915 N.E.2d 524, 534 (Ind. Ct. App. 2009).
16. *Id.*
17. *Id.* at 534–35.

Cinergy illustrates that not all emissions-related losses result from an occurrence. Consequently, insurers should carefully examine whether losses claimed in climate change-related lawsuits, especially losses sought as the result of governmental regulation, are the result of an occurrence, within the meaning of the policy. Amounts spent to prevent future harm may not fall within the definition of damages *caused* by an occurrence.

> Prophylactic measures required by EPA and meant to prevent future GHG emissions likely are not damages caused by an accident within the meaning of the term "occurrence" and, therefore, likely are not covered under a CGL policy.

3. The Effect of the "Trigger" on Occurrence

In theory, coverage is triggered under a CGL occurrence-based policy when bodily injury or property damage occurs. In reality, determining when coverage is triggered is complex if an injury takes place over an extended period of time. Furthermore, because injuries arising out of environmental causes and climate change will often post-date the initial exposure to the injurious cause, courts often wrestle with which policy applies to an injury and when coverage for that injury was triggered.[18] Climate change claims will pose even further unique challenges in comparison to other environmental claims as the harm allegedly caused by climate change is not localized but instead occurs globally, crossing national and international borders.

In an attempt to remove uncertainty regarding when coverage is triggered, courts, insureds, and insurers have insisted on developing global coverage triggers based on broad assumptions about when injury or damage occurs, which apply to all claims, irrespective of how different the claims may be. While the goal of developing such global triggers was to make determining when coverage is triggered more uniform, in reality the result has been referred to as "conceptual chaos."[19] From courts' attempts to determine which policy period is triggered, the following four principal trigger-of-coverage theories have developed: exposure, manifestation, injury-in-fact, and continuous. Each trigger is discussed below. Insurers should be aware that some states

---

18. With respect to most claims asserting climate change or some environmental cause as the source of injury, exposure to the injurious cause may occur during and continue through numerous policy periods, the resulting injury may develop during subsequent policy periods, and the manifestation of that injury may occur in yet another policy period. Keene Corp. v. Ins. Co. of N. Am., 667 F.2d 1034, 1040 (D.C. Cir. 1981).

19. Am. Home Prods. Corp. v. Liberty Mut. Ins. Co., 565 F. Supp. 1485, 1511 (S.D.N.Y. 1983), *aff'd as modified*, 748 F.2d 760 (2d Cir. 1984).

apply only one trigger, regardless of the type of claim, and others may apply different theories depending on the type of claim. Thus, which trigger will apply to which claim will be jurisdiction specific and sometimes even judge specific.

> Whether a policy is "triggered" will depend on the jurisdiction in which the action is pending.

a. Exposure[20]

Under the exposure trigger theory, injury occurs at the same time the injured party is exposed to the injury-causing agent. Thus, under the exposure trigger, each time a party comes into contact with the injury-causing agent constitutes an occurrence, and each insurer on the risk at the time of exposure should be liable for coverage with total liability prorated among all of the insurers.[21]

Applying this theory to claims arising out of injury due to climate change, an injury would be deemed to occur at the time a party is initially exposed to the GHG, and subsequent exposures occur every time the party comes into contact with the GHG. Thus, multiple policies will probably be implicated should a plaintiff be successful in proving a climate change-related injury.[22]

b. Manifestation[23]

Under the manifestation trigger theory, bodily injury or property damage is not deemed to have occurred, and coverage is not triggered, until the injury or damage resulting from the injury-causing agent manifests.[24] The date of manifestation is the date the injured party knew or should have known that he or she had suffered an

---

20. Alabama, Alaska, Arizona, California, Colorado, Connecticut, Georgia, Illinois, Louisiana, Maryland, Massachusetts, Michigan, Minnesota, Missouri, New Hampshire, New Jersey, New Mexico, New York, Ohio, Pennsylvania, Texas, Vermont, Washington, Wisconsin, Wyoming, and Wisconsin are states that have applied the exposure trigger theory. *See* STEVEN PLITT & JORDAN R. PLITT, 2 PRACTICAL TOOLS FOR HANDLING INSURANCE CASES § 13.5 (2011).

21. This theory, along with the three other trigger of coverage theories, arose primarily out of the glut of asbestos litigation in the 1970s and 1980s. The exposure theory has its origins in medical data showing that injury stemming from asbestos inhalation initially occurred, at least on a subcellular level, at the time the injured party was exposed to asbestos. Ins. Co. of N. Am. v. Forty-Eight Insulations, Inc., 633 F.2d 1212, 1217 (6th Cir. 1980).

22. To show causation, plaintiffs will likely need to prove market theory causation and that GHGs are fungible. Whether a court will accept such proof of causation is yet undetermined.

23. Arizona, California, District of Columbia, Georgia, Illinois, Indiana, Louisiana, Maine, Maryland, Massachusetts, Michigan, Minnesota, Missouri, Nevada, New Hampshire, New Jersey, New York, North Carolina, Ohio, Oklahoma, Pennsylvania, Rhode Island, South Carolina, Texas, Virginia, and Wisconsin are states that have applied the manifestation trigger theory. *See* PLITT & PLITT, *supra* note 20, § 13.5.

24. *Forty-Eight Insulations*, 633 F.2d at 1216.

injury.[25] Accordingly, when the injury is discovered, only the policy in effect will be triggered. Proponents of the manifestation theory assert that the application of this trigger of coverage theory is supported by the ordinary meaning of the terms "bodily injury, sickness or disease," as this phrase is generally used to define an occurrence.[26]

The difficulty of applying a manifestation trigger with respect to climate-change claims will depend on the type of claim. For instance, it may be hard to prove when the *Kivalina* plaintiffs knew, or should have known of the erosion to the Kivalina coastline. By contrast, in cases like *Comer* where there is a defined episodic event, the injury will be easier to determine if it allegedly resulted from a defined episodic event—a hurricane.

c. Injury-in-Fact[27]

The injury-in-fact trigger theory is broader than both the exposure and manifestation theories and has been adopted by an increasing number of courts because it is deemed fair to both the insured and the insurer. Under this theory, when the injury occurs is determined by the facts of each case and depends on proof that actual injury occurred during the policy period. Thus, coverage is triggered without regard to the time of exposure or when the injury actually manifested itself, even if both of those events took place outside of the policy period.

This theory was first advanced by the Southern District of New York in *American Home Products Corp. v. Liberty Mutual Insurance Co.*[28] In *American Home Products*, the court stated that when injury occurs is a function of the facts of each case, the injury-causing agent involved, the period and intensity of exposure, and the person affected. In light of these factors, an injury may occur upon exposure, at some point in time after exposure but before manifestation of the injury, or at manifestation.[29] "So long as the insured is held liable for an identifiable and compensable injury, sickness, or disease that is shown to have existed during coverage, that liability will be insured against whether or not the injury coincides with exposure or manifestation."[30] The court pointed to the policy's language and background,[31] the intentions and

---

25. Eagle-Picher Indus., Inc. v. Liberty Mut. Ins. Co., 682 F.2d 12, 16 (1st Cir. 1982).
26. *Keene*, 667 F.2d at 1043.
27. Alabama, Arizona, California, Colorado, Connecticut, Florida, Georgia, Idaho, Illinois, Indiana, Kansas, Maryland, Massachusetts, Michigan, Minnesota, Missouri, Montana, New Hampshire, New Jersey, New York, North Carolina, Oklahoma, Oregon, Pennsylvania, South Carolina, Texas, Utah, and Washington are states that have applied the injury-in-fact trigger theory. *See* Plitt & Plitt, *supra* note 20, § 13.5.
28. 565 F. Supp. 1485 (1983).
29. *Id.* at 1490.
30. *Id.* at 1498.
31. The policy at issue contained a standard occurrence-based liability coverage provision that provided coverage for occurrences resulting from personal injury, sickness or disease, including death resulting therefrom, sustained by any person. *Id.* at 1489.

expectations of the parties, and considerations of practicability and fairness as supporting this theory. Further, the injury-in-fact trigger of coverage provides liberal protection for the insured without running afoul of the principle that insurance policies are contracts under which insureds obtain all the protection for which they may reasonably be deemed to have paid, but not more.[32]

In cases where the injury accumulates, such as in the *Kivalina* case, the injury-in-fact trigger may be difficult to apply because of the challenges in determining when a harmful condition has accumulated to such a degree as to have caused an identifiable and compensable injury. Thus, this trigger may not be used in connection with claims relating to the cumulative effect of GHG emissions.

d. Continuous[33]

The continuous trigger, or "triple trigger," theory is the most expansive trigger of coverage theory and, accordingly, is the theory most favorable to insureds. The premise of this theory is that any insurance policy on the risk *at any point in time* between the claimant's initial exposure to the injury-causing agent and the manifestation of the resulting injury is responsible for coverage.[34] Under the continuous trigger theory, deteriorating damage or injury is recognized as multiple, continuously repeating occurrences along a timeline from when damage first appears until it terminates. Thus, the continuous or deteriorating damage is eligible for coverage under every policy in existence during the period that damage occurred. Under the continuous trigger theory, each policy covers its share of damage that occurred during the time the policy was in effect.[35]

The continuous trigger theory is congruent with the broad language of most CGL occurrence-based policies, which generally define an "occurrence" as the initial accident *as well as* repeated exposures to the injury-causing agent. Because it is often impossible to identify the exact point at which an injury resulting from climate change or environmental causes was sustained, an insured can seek coverage for a company's entire history of emissions, from the time the company first emitted a pollutant until the time the resulting damage was deemed to cause environmental injury.

---

32. *Id.*
33. Arizona, California, Colorado, Delaware, District of Columbia, Georgia, Hawaii, Idaho, Illinois, Indiana, Kansas, Maryland, Massachusetts, Michigan, Minnesota, Missouri, New Hampshire, New Jersey, New York, Ohio, Oregon, Pennsylvania, South Carolina, Texas, Washington, West Virginia, and Wisconsin are states that have applied the continuous trigger theory in certain circumstances. *See* PLITT & PLITT, *supra* note 20, § 13.5.
34. *Keene*, 667 F.2d at 1042.
35. *See, e.g.*, Montrose Chem. Corp. v. Admiral Ins. Co., 913 P.2d 878 (Cal. 1995).

Because the damage caused by GHG emissions may be considered latent and continuous, insurers and insureds should closely evaluate application of the continuous trigger theory in the climate change context. In particular, insurers and insureds should consider whether "stacking" of multiple policies could arise.

4. Long-Tail Claims

Depending on the trigger theory applied, progressive or continuous injury or damage claims, such as in the *Kivalina* case, could possibly trigger insurance coverage under multiple historical insurance policies purchased many years ago. These so-called long-tail claims pose unique coverage questions to insurers and insureds because they often implicate many years of coverage in which the insured has typically procured varying amounts and levels of primary and excess liability coverage from multiple insurers. Assuming such claims are even covered, problems may arise regarding how the loss or losses will be distributed or allocated among the various policy years and levels of coverage. This allocation will depend on resolution of a number of issues, including the trigger of coverage, number of occurrences, anti-stacking of limits provisions, inclusion of batch clauses, the court's allocation method for damages (pro-rata versus joint and several), and exhaustion method (horizontal versus vertical), among others.

C. Pollution Exclusions and Grants of Pollution Coverage

The interpretation of the term "pollutant" is likely to be highly relevant to climate change exposure. If GHGs are determined to be "pollutants" for purposes of insurance policies, CGL and certain D&O exposures may be reduced due to operation of the pollution exclusion, and environmental liability exposure may increase. Alternatively, a court construing an exclusion narrowly and a coverage grant broadly could interpret the term pollutant in a way that exposes insurers on both CGL and environmental liability books of business (i.e., stacking). As explained in further detail below, the law interpreting the term "pollutant," in the context of pollution exclusions and coverage grants, is rather unsettled and jurisdiction specific.

1. Definition of "Pollutant"

> The definition of the term "pollutant" is likely to be highly relevant to climate change exposure.

Although definitions vary by policy, a "pollutant" is typically defined as "any solid, liquid, gaseous, or thermal irritant or contaminant, including smoke, vapor, soot, fumes, acids, alkalis, chemicals and waste. Waste includes materials to be recycled, reconditioned or reclaimed."[36] Some courts have interpreted the term "pollutant" broadly to encompass many pollutants.[37] Other courts have limited pollution exclusions to specific situations involving "traditional" environmental pollution and pollutants because they find the term pollutant as used in pollution exclusions to be ambiguous and believe that policyholders have reasonable expectations of coverage for claims involving "non-traditional" pollution.[38] The definition of "pollutant" is policy specific and jurisdiction specific, and thus the specific policy language and law of the jurisdiction at issue probably will impact the ultimate treatment of GHGs. Below is a discussion of (1) historical litigation involving the term "pollutant" and how such litigation may act as a guide to whether GHGs will be treated as pollutants and (2) recent and ongoing litigation regarding whether GHGs are pollutants for purposes of a pollution exclusion.

a. The Interpretation of the Term "Pollutant"

Whether GHGs will be considered "pollutants" will depend on the language in the policy and the jurisdiction in which the litigation is brought.

Whether a substance qualifies as a pollutant for purposes of a pollution exclusion is jurisdiction specific and heavily influenced by the characterization of pollution exclusion clauses as ambiguous or unambiguous. For instance, courts deeming such clauses to be unambiguous have held that substances such as carbon monoxide,[39] nitrogen dioxide (a GHG),[40] hydrogen sulfide,[41] styrene,[42] and anhydrous ammonia[43] qualify as pollutants and therefore fall within the scope of a pollution exclusion. In each of the referenced cases, the court reasoned that the substance was a "solid, liquid, gaseous or thermal irritant or contaminant" and that no further analysis was required.[44] Courts in these jurisdictions likewise may conclude that GHGs are pollutants because they arguably are gaseous contaminants.

---

36. ISO Form CG 00 01 12 07.
37. *See* Whittier Props., Inc. v. Alaska Nat'l Ins. Co., 185 P.3d 84, 89–92 (Alaska 2008); Heyman Assocs. No. 1 v. Ins. Co. of Pa., 653 A.2d 122, 129–33 (Conn. 1995); Deni Assocs. of Fla., Inc. v. State Farm Fire & Cas. Ins. Co., 711 So. 2d 1135, 1137–41 (Fla. 1998); Reed v. Auto-Owners Ins. Co., 667 S.E.2d 90, 92 (Ga. 2008); Bituminous Cas. Corp. v. Sand Livestock Sys., Inc., 728 N.W.2d 216, 220–22 (Iowa 2007).
38. *See* Keggi v. Northbrook Prop. & Cas. Ins. Co., 13 P.3d 785, 790–92 (Ariz. Ct. App. 2000); Minerva Enters., Inc. v. Bituminous Cas. Corp., 851 S.W.2d 403, 404–06 (Ark. 1993); MacKinnon v. Truck Ins. Exch., 73 P.3d 1205, 1208–18 (Cal. 2003); Am. States Ins. Co. v. Koloms, 687 N.E.2d 72, 75–82 (Ill. 1997); Am. States Ins. Co. v. Kiger, 662 N.E.2d 945, 948–49 (Ind. 1996).
39. Bernhardt v. Hartford Fire Ins. Co., 648 A.2d 1047 (Md. Ct. Spec. App. 1994).
40. League of Minn. Cities Ins. Trust v. City of Coon Rapids, 446 N.W.2d 419 (Minn. Ct. App. 1989).
41. United Nat'l Ins. Co. v. Hydro Tank, Inc., 497 F.3d 445 (5th Cir. 2007).
42. Hydro Sys., Inc. v. Cont'l Ins. Co., 717 F. Supp. 700 (C.D. Cal. 1989).
43. TerraMatrix, Inc. v. U.S. Fire Ins. Co., 939 P.2d 483, 487–88 (Colo. Ct. App. 1997).
44. *See, e.g., Bernhardt*, 648 A.2d at 1051.

When courts have found pollution exclusions to be ambiguous, some of the same substances addressed in the preceding paragraph have been deemed not to fall within the scope of a pollution exclusion.[45] It is less clear from this precedent, however, how these jurisdictions would treat GHGs because, often, the basis for each court's determination that a substance is not a pollutant would not apply to GHGs. For example, some courts have refused to designate a substance a pollutant based on the containment of that substance in the area of its intended use as part of the insureds' normal business activities.[46] GHGs alleged to cause climate change would not meet that criterion. In addition, many of these cases have involved smaller claimants. Some courts may be more likely to conclude that a gas or fume falls within the pollution exclusion in an industrial versus a homeowner or landlord-tenant situation.

How a jurisdiction treats commonly occurring substances with regard to a pollution exclusion may prove indicative of how GHGs, such as water vapor and CO_2, will be treated in that jurisdiction. Jurisdictions deeming exclusions to be unambiguous and applying the terms of an exclusion literally have had no difficulty designating commonly occurring substances as pollutants.[47] For example, a Minnesota appeals court held that gases and fumes from a pig farm fell within the pollution exclusion because the pollution exclusion clearly applied to fumes.[48] In *Colony National Insurance Co. v. Specialty Trailer Leasing, Inc.*, a district court in Texas held that argon, despite being an inert, naturally occurring gas, could constitute a pollutant at "dangerously elevated concentrations."[49]

In contrast, relying on the principle that insureds have reasonable expectations of coverage for damages caused by substances that are not traditionally considered to be pollutants,[50] some jurisdictions deeming exclusions to be ambiguous have found

---

45. *See* Donaldson v. Urban Land Interests, Inc., 564 N.W.2d 728, 732–33 (Wis. 1997) (holding that exhaled CO_2 is not a pollutant for purposes of interpreting a pollution exclusion because a reasonable insured may not understand that personal injury claims having their genesis in activities as fundamental as human respiration would be excluded from coverage).

46. *See, e.g.*, Essex Ins. Co., Inc. v. Berkshire Envtl. Consultants, Inc., No. Civ. A. 99-30280-FHF, 2002 WL 226172, at *3 (D. Mass. Feb. 7, 2002) (holding that hydrogen sulfide confined to the area of its intended use is not a "pollutant").

47. In fact, on at least one occasion, a court has expressly rejected the argument that a substance must generally act as an irritant or contaminant to be considered a pollutant. *See* Nautilus Ins. Co. v. Country Oaks Apts. Ltd., 566 F.3d 452, 455 (5th Cir. 2009).

48. Wakefield Pork, Inc. v. Ram Mut. Ins. Co., 731 N.W.2d 154, 160 (Minn. Ct. App. 2007); *see also* Cont'l Cas. Co. v. Advance Terrazzo & Tile Co., 462 F.3d 1002, 1008–09 (8th Cir. 2006) (applying Minnesota law and holding that carbon monoxide released as a byproduct of industrial grinders was a pollutant for purposes of the CGL pollution exclusion clause).

49. 620 F. Supp. 2d 786, 790 (N.D. Tex. 2009).

50. Though the concept of a "traditional environmental pollutant" or "traditional environmental pollution" is somewhat amorphous, courts appear to apply a "reasonable person" standard to these terms. For example, in *Donaldson v. Urban Land Interests, Inc.*, 564 N.W.2d 728, the Wisconsin Supreme Court held that exhaled CO_2

that commonly occurring gases cannot qualify as pollutants under a pollution exclusion clause. For example, an appeals court in Wisconsin held that carbon monoxide at high levels in a residence resulting from operation of a fireplace and boiler was not a pollutant within the meaning of the landlord's liability policy partially because the landlord could reasonably expect coverage for damages caused by the accumulation of a substance that is commonly present.[51] In these types of jurisdictions, insureds seeking to negate application of a pollution exclusion may argue that GHGs (in particular, CO_2) are commonly occurring, inert gases typically found in the atmosphere, and as such do not qualify as traditional pollutants for purposes of the pollution exclusion.

b. Whether GHGs Are Pollutants
(1) *Massachusetts v. EPA* and Recent EPA Regulatory Actions

> The Supreme Court considers certain GHGs to be pollutants for purposes of the Clean Air Act. Such a conclusion arguably provides persuasive authority that those GHGs should be considered pollutants for insurance purposes.

An important development regarding whether courts will recognize GHGs as pollutants for purposes of pollution exclusion clauses is the 2007 U.S. Supreme Court decision in *Massachusetts v. EPA*.[52] In *Massachusetts*, the Supreme Court held that certain GHGs (CO_2, CH_4, N_2O, and HFCs) are pollutants for purposes of the Clean Air Act. See Chapter 5.B.1.a detailing the decision.

In response to *Massachusetts v. EPA*, EPA has taken several actions that provide further support for the contention that GHGs are pollutants—at least currently. EPA has finalized a reporting rule pursuant to the Clean Air Act requiring high-emitting facilities to submit to EPA annual reports detailing emissions of CO_2, CH_4, N_2O, SF_6, HFCs, PFCs, and other fluorinated gases without stating whether those gases qualify as pollutants (see Chapter 5.B.4); issued an Endangerment and Cause or Contribute Finding defining "air pollution" under the Clean Air Act to be the "mix" of CO_2, CH_4, N_2O, HFCs, PFCs, and SF_6 (see Chapter 5.B.1.b); issued a rule creating light-duty vehicle emission standards for air pollutants CO_2, CH_4, N_2O, and HFCs under the Clean Air Act (see Chapter 5.B.2); issued a rule tailoring certain Clean Air Act

---

is not a pollutant for purposes of interpreting a pollution exclusion because an activity as fundamental as human respiration would not be recognized by an insured as "traditional environmental pollution."

51. Langone v. Am. Family Mut. Ins. Co., 731 N.W.2d 334, 338–40 (Wis. Ct. App. 2007); *see also* Am. States Ins. Co. v. Koloms, 687 N.E.2d 72 (Ill. 1997).

52. 549 U.S. 497, 528–29 (2007). Other Clean Air Act cases following *Massachusetts* have further solidified the notion that CO_2, at least, is a pollutant under the Clean Air Act. *See In re* Deseret Power Elec. Coop., PSD Appeal No. 07-03, 2008 WL 5572891 (EAB Nov. 13, 2008); Longleaf Energy Assocs., LLC v. Friends of the Chattahoochee, Inc., 681 S.E.2d 203 (Ga. Ct. App. 2009).

emission thresholds for CO_2, CH_4, N_2O, HFCs, PFCs, and SF_6 (see Chapter 5.B.5); and acknowledged that CO_2, CH_4, N_2O, and HFCs would become pollutants "subject to regulation" under the Clean Air Act upon finalization of the vehicle emissions standards discussed above (see Chapter 5.B.3).

Thus, *Massachusetts* and these EPA rulings arguably provide persuasive evidence that GHGs should be treated as pollutants for insurance purposes. They are not, however, dispositive. The meaning of the term "pollutant" in the Clean Air Act context[53] may differ from insurance policy definitions of "pollutant." In addition, these decisions are recent and insureds may argue that they should not be used to construe pollution exclusions in occurrence-based policies issued before *Massachusetts* and the enactment of referenced EPA regulations. In the environmental liability insurance context, however, insureds likely will rely on EPA's rulemaking to bolster arguments favoring coverage.

> EPA's GHG Reporting Rule, Endangerment Finding, Cause or Contribute Finding, standards for light-duty vehicles, and tailoring rule may provide support for a finding that GHGs are pollutants for insurance purposes.

(2) *AES Corp. v. Steadfast Insurance Co.*

In *AES Corp. v. Steadfast Insurance Co.*, Steadfast sought a declaratory judgment that it had no duty to defend AES in the underlying *Kivalina* litigation[54] on three grounds including that coverage was barred by the pollution exclusion.[55] In issuing its final decision on a petition for rehearing, the Virginia Supreme Court did not reach the issue of whether GHGs fall within the definition of "pollutant" for purposes of the pollution exclusion because it held there was no occurrence.[56] Nevertheless, the arguments made by the parties in the course of the *AES* litigation with respect to the pollution exclusion issue provide an illustration of the types of arguments that may be made in future coverage litigation involving climate change-related claims.

As background, the pollution exclusion in the CGL policies at issue in *AES* provided that the policies do not apply to claims for damages "arising out of the actual, alleged or threatened discharge, dispersal, release or escape of pollutants."[57] The

---

53. The Clean Air Act defines "air pollutant" to include "any air pollution agent or combination of such agents, including any physical, chemical, biological, radioactive . . . substance or matter which is emitted into or otherwise enters the ambient air." Clean Air Act § 302(g), 42 U.S.C. § 7602(g)(2012).
54. *See* Chapter 6.A.3.
55. See Sections B.1 (occurrence) and D.3 (loss in progress) for a discussion of the two other grounds.
56. AES Corp. v. Steadfast Ins. Co., 725 S.E.2d 532 (Va. 2012).
57. Brief of Appellee at 5, *AES Corp.*, No. 100764 (Va. Oct. 8, 2010).

policies defined "pollutants" as "any solid, liquid, gaseous or thermal irritant or contaminant, including smoke, vapors, soot, fumes, acids, alkalis, chemicals and waste."[58]

On appeal to the Virginia Supreme Court, AES argued that CO_2 was not an "irritant" or a "contaminant."[59] Rather, CO_2 is an odorless and colorless gas emitted in the atmosphere in massive quantities by man-made and natural sources.[60] AES pointed out that, like H_2O, CO_2 is omnipresent in the environment.[61] Alternatively, AES argued that there are questions of fact regarding whether the parties considered CO_2 to be a pollutant.[62] Thus, according to AES, Steadfast cannot prove that "pollutant" unambiguously includes CO_2.

On appeal, Steadfast argued that while debate surrounds interpretation of the pollution exclusion in odd cases, there is no question that it applies to claims for environmental pollution.[63] Steadfast argued that the *Kivalina* complaint is "quintessentially" a claim alleging environmental pollution because it alleges that AES has released huge amounts of gaseous waste products and expressly characterizes the environmental damage as pollution.[64] Steadfast contended that under Virginia law, the characterization of the damages in the underlying case is dispositive.[65] The *Kivalina* plaintiffs characterized the emissions in question as pollution.[66]

Both parties claimed that the other relied on evidence outside the four corners of the policy to establish the meaning of the term "pollutant."[67] AES claimed that Steadfast could not look outside the four corners to federal statutes and cases and Virginia statutes.[68] Steadfast claimed that AES went to great lengths to cite deposition testimony of AES and insurance employees regarding their impressions of the term "pollutant."[69] Regardless, AES argued that neither EPA nor Virginia regulated CO_2 emissions from power plants at the time of contracting.[70]

The Virginia Supreme Court did not reach the issue of whether CO_2 is a pollutant during rehearing. The interpretation of the pollution exclusion in the climate change context may be the source of debate in future years.

---

58. *Id.* at 5–6.
59. *Id.* at 19.
60. *Id.*
61. *Id.* at 20.
62. *Id.*
63. Brief of Appellee at 38, *AES Corp.*, No. 100764 (Va. Sept. 13, 2010).
64. *Id.* at 38.
65. *Id.* at 39 (citing City of Chesapeake v. States Self-Insurers Risk Retention Grp., Inc., 628 S.E.2d 539, 541 (Va. 2006)).
66. *Id.* at 39–42.
67. Brief of Appellant at 22, *AES Corp.*, No. 100764 (Va. Oct. 8, 2010); Brief of Appellee, *supra* note 63, at 39–40.
68. Brief of Appellant, *supra* note 67, at 22.
69. Brief of Appellee, *supra* note 63, at 38.
70. Brief of Appellant, *supra* note 67, at 22.

> Whether GHGs are pollutants was one subject of the *AES* litigation.

(3) The Insureds' Conundrum: Whether to Argue that GHGs Are Pollutants

As various types of claims emerge, insureds may find that the decision of whether to argue that GHGs are pollutants is a difficult and context-specific one. That is because the question of whether GHGs are pollutants may arise in the context of interpreting both a coverage grant and a pollution exclusion. If insureds attempt to obtain coverage for GHG emissions under policies that specifically cover environmental liabilities they may find themselves in the awkward position of arguing that GHGs constitute pollutants for purposes of such policies while at the same time asserting that GHGs do not constitute pollutants for purposes of the pollution exclusion in CGL or other types of policies. The insured also may be debating how to treat classification of GHGs in various regulatory contexts. For example, in the context of CCS, some large utility companies have taken the position that CO_2 should not qualify as a hazardous waste for purposes of CERCLA and Resource Recovery and Conservation Act (RCRA). These utilities may find it difficult to then argue that CO_2 is a pollutant for purposes of recovering under an environmental insurance policy. A court, however, in construing a coverage grant broadly and an exclusion narrowly, theoretically could find no conflict and conclude that a pollution exclusion does not bar CGL coverage for climate change-related damage, while at the same time holding that an environmental policy provides coverage for such damages.

2. Pollution Exclusion

Pollution exclusions, which are standard clauses commonly found in CGL policies, may be highly relevant in the climate-change coverage context. Whether GHGs are pollutants is one key issue. The form of the exclusion is also important. There are three primary forms of pollution exclusion clauses in CGL policies (qualified, absolute, and total), although variations do exist. The qualified pollution exclusion began appearing in CGL policies in the 1970s,[71] the updated absolute pollution exclusion first appeared in the mid-1980s,[72] and the total pollution exclusion came into use in approximately 1999.[73] Given that insureds may look to historic occurrence-based CGL policies for defense and indemnification in the climate change context, all three are relevant.

---

71. *See, e.g.*, STEVEN PLITT ET AL., 9 COUCH ON INSURANCE § 127:10 (3d ed. 2011) [hereinafter COUCH ON INSURANCE].
72. *See, e.g., id.* § 127:13.
73. ISO Form CG 21 49 09 99.

Below is a discussion of the similarities and differences among the qualified (providing coverage for sudden and accidental releases), absolute (providing coverage for products and completed operations), and total (providing no coverage for any pollution events) pollution exclusions and the effect those differences can have on an insurer's obligations to its insureds. The discussions show that there is much legal uncertainty in how courts interpret the exclusions and any interpretation probably will be jurisdiction specific. The definition of the term "pollutant," discussed above in Section C.1, is also central to whether a "pollution exclusion" will act to bar coverage. The use of limited pollution coverage endorsements may affect the application of pollution exclusions as they provide coverage for short-term pollution events in certain circumstances. Finally, pollution exclusions may operate differently outside the CGL context, such as in D&O policies.

a. The Three Types of Pollution Exclusions
(1) The Qualified Pollution Exclusion

> The qualified pollution exclusion is found in historic policies and does not apply to discharges, dispersals, or releases that are "sudden and accidental." The effectiveness of this exclusion in barring coverage for GHG emissions will be jurisdiction specific, as some jurisdictions refuse to enforce the exclusion and others hold that "sudden and accidental" can encompass gradual releases.

The qualified pollution exclusion is relevant with respect to historic occurrence-based CGL policies that could still create liability for insurers. The qualified pollution exclusion with "sudden and accidental" language was first included in the 1973 version of ISO's general liability policy. It was not until the 1986 that ISO removed the "sudden and accidental" language, and stated that the exclusion applied to pollutant releases, whether "gradual or sudden." See Section C.2.a.(2), on the absolute pollution exclusion. Thus, general liability policies written before that time may contain the qualified pollution exclusion.

The qualified pollution exclusion typically provides that the insurance policy shall not apply

> to bodily injury or property damage arising out of the discharge, dispersal, release or escape of smoke, vapors, soot, fumes, acids, alkalis, toxic chemicals, liquids or gases, waste materials or other irritants, contaminants or pollutants into or upon land, the atmosphere or any water course or body of water; but this exclusion does not apply if such discharge, dispersal, release or escape is sudden and accidental.[74]

---

74. ISO Form GL 00 02 01 73, Exclusion (f).

Initially, the effectiveness of the qualified exclusion was somewhat limited as courts struggled with the meaning of the phrase "sudden and accidental." Some courts deemed the phrase ambiguous and therefore refused to enforce the exclusion.[75] Others interpreted "sudden and accidental" to encompass gradual releases, such that almost all claims were covered.[76] Still others enforced the language more narrowly, excluding all gradual releases from coverage.[77]

Insurers may attempt to argue that the sudden or accidental discharge requirement defeats coverage in the climate change context because GHG emissions are neither sudden nor accidental. GHG emissions generally are continuous and result from a business's normal operations. Indeed, many industries, including but not limited to oil, gas, chemical, agriculture, and coal have been emitting GHGs for many years. In jurisdictions that have held that damages caused by gradual pollution are not covered by policies with a qualified pollution exclusion because the pollution was not sudden or accidental,[78] an insurer could potentially succeed in arguing that the qualified pollution exclusion bars coverage.

(2) The Absolute Pollution Exclusion

> Whether a climate change-related claim will be excluded by the absolute pollution exclusion will depend on whether GHGs are considered pollutants within the meaning of the policy.

After much litigation surrounding the meaning of the "qualified pollution exclusion," the insurance industry attempted to negate the uncertainty created by the "sudden and accidental" language by drafting a new pollution exclusion, commonly referred to as the "absolute pollution exclusion." The absolute pollution exclusion, which first appeared in CGL policies in 1985, eliminates the exception for sudden and accidental releases. One typical absolute pollution exclusion provides that CGL coverage will not apply to

(1) "Bodily injury" or "property damage" arising out of the actual, alleged or threatened discharge, dispersal, seepage, migration, release or escape of "pollutants":
   (a) At or from any premises, site or location which is or was at any time owned or occupied by, or rented or loaned to, any insured. However, this subparagraph does not apply to:

---

75. Am. States Ins. Co. v. Kiger, 662 N.E.2d 945 (Ind. 1996) (holding that pollution exclusion and "sudden and accidental" exception to exclusion in CGL policy were ambiguous and therefore did not preclude coverage for damages caused by leakage of underground gasoline storage tanks).
76. *See, e.g.,* Northville Indus. Corp. v. Nat'l Union Fire Ins. Co. of Pittsburgh, Pa., 679 N.E.2d 1044, 1047 (N.Y. 1997).
77. *See, e.g.,* Fireman's Fund Ins. Cos. v. Ex-Cell-O Corp., 702 F. Supp. 1317, 1326–27 (E.D. Mich. 1998).
78. *See, e.g.,* U.S. Fid. & Guar. Co. v. Star Fire Coals, Inc., 856 F.2d 31, 34–35 (6th Cir. 1988).

(i) "Bodily injury" if sustained within a building and caused by smoke, fumes, vapor or soot produced by or originating from equipment that is used to heat, cool or dehumidify the building, or equipment that is used to heat water for personal use, by the building's occupants or their guests;

(ii) "Bodily injury" or "property damage" for which you may be held liable, if you are a contractor and the owner or lessee of such premises, site or location has been added to your policy as an additional insured with respect to your ongoing operations performed for that additional insured at that premises, site or location and such premises, site or location is not and never was owned or occupied by, or rented or loaned to, any insured, other than that additional insured; or

(iii) "Bodily injury" or "property damage" arising out of heat, smoke or fumes from a "hostile fire",

(b) At or from any premises, site or location which is or was at any time used by or for any insured or others for the handling, storage, disposal, processing or treatment of waste;

(c) Which are or were at any time transported, handled, stored, treated, disposed of, or processed as waste by or for:
   (i) Any insured; or
   (ii) Any person or organization for whom you may be legally responsible.

(d) At or from any premises, site or location on which any insured or any contractors or subcontractors working directly or indirectly on any insured's behalf are performing operations if the "pollutants" are brought on or to the premises, site or location in connection with such operations by such insured, contractor or subcontractor. However, this subparagraph does not apply to:

(i) "Bodily injury" or "property damage" arising out of the escape of fuels, lubricants or other operating fluids which are needed to perform the normal electrical, hydraulic or mechanical functions necessary for the operation of "mobile equipment" or its parts, if such fuels, lubricants or other operating fluids escape from a vehicle part designed to hold, store or receive them. This exception does not apply if the "bodily injury" or "property damage" arises out of the intentional discharge, dispersal or release of the fuels, lubricants or other operating fluids, or if such fuels, lubricants or other operating fluids are brought on or to the premises, site or location with the intent that they be discharged, dispersed or released as part of the operations being performed by such insured, contractor or subcontractor;

(ii) "Bodily injury" or "property damage" sustained within a building and caused by the release of gases, fumes or vapors from materials brought into that building in connection with operations being performed by you or on your behalf by a contractor or subcontractor; or

(iii) "Bodily injury" or "property damage" arising out of heat, smoke or fumes from a "hostile fire".

(e) At or from any premises, site or location on which any insured or any contractors or subcontractors working directly or indirectly on any insured's behalf are performing operations if the operations are to test for, monitor, clean up, remove, contain, treat, detoxify or neutralize, or in any way respond to, or assess the effects of, "pollutants".

(2) Any loss, cost or expense arising out of any:
(a) Request, demand, order or statutory or regulatory requirement that any insured or others test for, monitor, clean up, remove, contain, treat, detoxify or neutralize, or in any way respond to, or assess the effects of, "pollutants"; or
(b) Claim or "suit" by or on behalf of a governmental authority for damages because of testing for, monitoring, cleaning up, removing, containing, treating, detoxifying or neutralizing, or in any way responding to, or assessing the effects of, "pollutants".

However, this paragraph does not apply to liability for damages because of "property damage" that the insured would have in the absence of such request, demand, order or statutory or regulatory requirement, or such claim or "suit" by or on behalf of a governmental authority.[79]

Unlike the qualified pollution exclusion, the majority of courts that have considered the absolute pollution exclusion have found it unambiguous, but that does not mean the exclusion always operates to bar coverage.[80] Courts refusing to apply the absolute exclusion typically have found that the substance at issue did not qualify as a pollutant pursuant to the terms of a given policy. As a result, the absolute pollution exclusion has not led to a decrease in litigation; the thrust of the litigation has merely evolved from interpreting "sudden and accidental" to interpreting the term "pollutant." Consequently, whether a climate change-related claim will be excluded will probably depend on whether the jurisdiction where coverage is being litigated holds that GHGs constitute "pollutants" within the meaning of the policy. See Section C.1 discussing how courts have interpreted what constitutes a "pollutant."

In addition, it bears noting that the absolute pollution exclusion contains many exceptions that must be taken into account during underwriting, as the exceptions in some circumstances may swallow the exclusion. For instance, a pollution event occurring from a mobile source (such as vehicle) is covered if the mobile source requires fuels, lubricants, or other operating fluids for normal operation. An insured could argue that GHG emissions from a vehicle do not fall under the pollution exclusion due to this exception. Also, if GHGs are emitted pursuant to a hostile fire, these emissions could also be excepted from the absolute pollution exclusion.

---

79. ISO Form, CG 00 01 12 07.
80. *See, e.g.*, 9 COUCH ON INSURANCE, *supra* note 71, § 127:14 (2011).

> Insurers must also be aware that the absolute pollution exclusion has many exceptions that potentially could swallow the exclusion in the climate change context.

(3) The Total Pollution Exclusion

The total pollution exclusion, unlike the absolute pollution, excludes coverage for products and completed operations and off-site releases.[81] The basic total pollution exclusion states:

> **COMMERCIAL GENERAL LIABILITY COVERAGE PART**
>
> Exclusion f. under Paragraph 2., Exclusions of Section I—Coverage A—Bodily Injury And Property Damage Liability is replaced by the following:
> This insurance does not apply to:
> f. Pollution
> (1) "Bodily injury" or "property damage" which would not have occurred in whole or part but for the actual, alleged or threatened discharge, dispersal, seepage, migration, release or escape of "pollutants" at any time.
> (2) Any loss, cost or expense arising out of any:
> (a) Request, demand, order or statutory or regulatory requirement that any insured or others test for, monitor, clean up, remove, contain, treat, detoxify or neutralize, or in any way respond to, or assess the effects of "pollutants"; or
> (b) Claim or suit by or on behalf of a governmental authority for damages because of testing for, monitoring, cleaning up, removing, containing, treating, detoxifying or neutralizing, or in any way responding to, or assessing the effects of, "pollutants".[82]

The total pollution exclusion is arguably the broadest, most far-reaching exclusion. This does not mean, however, that courts necessarily will enforce it as intended by the insurance industry.

There are few cases interpreting the total pollution exclusion and courts and legal experts routinely confuse the absolute and total pollution exclusions. Consequently, there is much uncertainty regarding how courts will apply the total pollution exclusion. Those courts that have analyzed the total pollution exclusion, however, have generally found the provision ambiguous in light of its broad exclusionary language. Some courts have held that a literal reading of the exclusion is "absurd" because it excludes coverage for any claim regarding any contaminant or irritant.[83] For instance,

---

81. *See, e.g.*, 9 Couch on Insurance, *supra* note 71, § 127:3 (2011).
82. ISO Total Pollution Exclusion Endorsement, CG 21 49 09 99.
83. *See, e.g.*, Doerr v. Mobil Oil Corp., 774 So. 2d 119, 124 (La. 2000); In re Idleaire Technologies Corp., Bankr. No. 08-10960 (KG), 2009 WL 413117, at *8–9 (Bankr. D. Del. Feb. 18, 2009).

at least one court has theorized that bodily injury caused by slipping on spilled Drano could fall under the total pollution exclusion because Drano can be considered an irritant.[84] Thus, some courts have limited the exclusion to environmental pollution.[85] Such courts have held that whether an event will constitute environmental pollution will turn upon whether the insured is a "polluter" within the meaning of the policy; whether the injury causing substance is a "pollutant" within the meaning of the policy; and whether there was a "discharge, dispersal, seepage, migration, release or escape" of a pollutant by the insured within the meaning of the policy.[86] See Section C.1 for a discussion of how courts have applied the definition of the term "pollutant."

Although courts have interpreted the total pollution exclusion more narrowly than the insurance industry may have intended, it is unlikely that the environmental pollution interpretation will have much effect on most climate change-related claims, such as those that were brought in *Kivalina*, *Comer*, *AEP*, and *Turner*. See Chapter 6.A. Arguably, the defendants in those cases could all be classified as traditional polluters. In addition, it would be hard for any of the defendants to argue that their emissions (or the oil spill in the case of *Turner*) do not constitute a discharge, dispersal, or seepage. Thus, like the absolute pollution exclusion, whether the total pollution exclusion will apply in climate change-related claims probably will turn upon whether GHGs are considered "pollutants" in a particular claim scenario.

If courts interpret the total pollution exclusion literally, the total pollution exclusion may reduce exposure to claims in which the absolute pollution exclusion would not have barred coverage, such as those where an automobile company seeks coverage related to a lawsuit alleging that the company manufactured automobiles that are defective as designed because they emit GHGs. The total pollution exclusion arguably excludes coverage for pollution generated from products (automobiles) and any pollution event no matter where it occurs (on the roads instead of on the manufacturer's premises). The absolute pollution exclusion, on the other hand, arguably does not exclude coverage for completed operations and/or for pollution events that do not occur on an insured's premises.[87] Because the total pollution exclusion has yet to be tested with respect to climate change-related claims and courts routinely confuse the absolute and total pollution exclusions, whether the total pollution exclusion can bar coverage for climate change-related claims remains uncertain.

---

84. *Doerr*, 774 So. 2d at 124.
85. *Id.* at 135.
86. *Id.*; *see also In re Idleaire*, 2009 WL 413117, at *9.
87. *See, e.g.*, 9 COUCH ON INSURANCE, *supra* note 71, § 127:3 (2011).

b. Time Element, Limited-Pollution Coverage Endorsements

In some circumstances, insurers may carve back the absolute or total pollution exclusions in a policy by granting limited pollution coverage endorsements. These endorsements provide coverage for certain short-term pollution events beginning and ending in a short time frame, like 48 or 72 hours. For example, ISO's Limited Exception for Short-Term Pollution Events, CG 04 29 02 02, defines "short-term pollution event" as a discharge, dispersal, release or escape of "pollutants" which:

a. Begins during the policy period;
b. Begins at an identified time and place;
c. Ends, in its entirety, at an identified time within forty-eight (48) hours of the beginning of the discharge, dispersal, release or escape of the "pollutants";
d. Is not a repeat or resumption of a previous discharge, dispersal, release or escape of the same pollutant from essentially the same source within twelve (12) months of a previous discharge, dispersal, release or escape;
e. Does not originate from an "underground storage tank"; and
f. Is not heat, smoke or fumes from a "hostile fire."

To be a "short-term pollution event," the discharge, dispersal, release or escape of "pollutants" need not be continuous. However, if the discharge, dispersal, release or escape is not continuous, then all discharges, dispersals, releases or escapes of the same "pollutants" from essentially the same source, considered together, must satisfy Provisions a. through f. of this definition to be considered a "short-term pollution event."[88]

Given the short time periods associated with these endorsements, potential exposure in the climate change context would most likely occur with respect to a *Turner v. Murphy Oil*–type situation in which a temporary, distinct episodic event is involved. See Chapter 6.A.5.a. Oil spills and other potentially short-term environmental disasters arising out of episodic climatic events might fit within the terms of these endorsements, depending on the particular facts of the claim. It is less likely that an insured would attempt to argue or succeed in arguing that limited pollution endorsements provide coverage for *Kivalina*-type claims given that alleged emissions did not begin and end over a 48- or 72-hour period. Transactional costs and defense costs, however, could still be significant if someone attempted to make such a case.

It should also be noted that although "short-term pollution event" is often defined as either 48 or 72 hours, insurers may offer extended time frames of days or even weeks. Such extensions of the time period would provide insureds with much greater coverage than standard endorsements. In addition, coverage may further be extended by courts. At least one court has interpreted the meaning of "commencement of a

---

88. ISO Limited Exception for Short-Term Pollution Events, CG 04 29 02 02.

pollution event" in a limited pollution coverage endorsement to mean the time when the pollution event is reasonably discoverable.[89] Consequently, it is possible that a pollution event, like a spill or leak, could occur for months or even years and still be considered "short term" if the insured could not reasonably discover the event during that time and once discovered, the insured promptly, within the time frame provided by the policy, causes the pollution event to end.

> Limited pollution coverage endorsements may provide coverage for short-term episodic climatic events like *Turner* but most likely will not have a large impact on tort claims such as those found in *Kivalina* and *Comer*.

c. Special Issues Associated with Pollution Exclusions in the D&O Context

The risks related to climate change are not limited to CGL policies. It is likely that D&O policies also will be implicated. In the D&O context, shareholders and regulators could decide to sue directors and officers for mismanagement of climate change risks and failure to adequately disclose any risks. For a discussion of the climate change-related risks associated with D&O policies and background on climate change disclosure issues, see Chapter 7.A.3.a and Chapter 6.E.2–5, respectively.

As with CGL coverage cases related to climate change, D&O coverage cases may involve interpretation of whether the pollution exclusion applies to climate change claims. Most significantly, courts will need to consider whether GHGs are pollutants within the meaning of the pollution exclusion. In the case of D&O policies, however, courts also may have to determine whether pollution exclusions apply to claims arising out of misrepresentation or mismanagement of climate change-related risks as opposed to claims arising more directly from pollution. Some D&O policies have pollution exclusions with broad "arising out of" language. In interpreting such clauses, it appears that courts will focus on the "causal link" between pollution and harm.

Thus far, courts have reached different results when faced with this issue. In *Sealed Air Corp. v. Royal Indemnity Co.*,[90] a New Jersey appellate court considered whether a pollution exclusion barred coverage. In the underlying action, plaintiffs had alleged that Sealed Air's directors and officers caused the company's stock to trade at artificially high prices by issuing false and misleading statements regarding contingent asbestos claims asserted against the company for past acts of pollution.[91] The court concluded that the alleged loss asserted by the company's shareholders arose not from

---

89. Vermont Mut. Ins. Co. v. Am. Home Assurance Co., Nos. WC04-0591, WC08-0137, slip op. at *8 (R.I. Super. Ct. Aug. 4, 2009).
90. 961 A.2d 1195, 1201 (N.J. Super. Ct. App. Div. 2008).
91. *Id.* at 1203.

any underlying pollution, but from the allegedly misleading financial statements. The court instructed that the pollution exclusion requires a more direct causal relationship between the pollution and the harm asserted. Thus, the court held that the pollution exclusion did not apply to bar coverage.[92]

Similarly, in *Owens Corning v. National Union Fire Insurance Co. of Pittsburgh, Pa.*,[93] the Sixth Circuit held that an asbestos pollution exclusion did not bar coverage in a class action by shareholders alleging misrepresentation of financial exposure to asbestos claims and sufficiency of insurance coverage.[94] As in *Sealed Air*, the court noted that the shareholders' allegations were not based on Owens Corning's use of asbestos, but on the company's deceit of investors, which had no discernible connection to product-liability issues such as those covered by the asbestos exclusion.[95]

In contrast, in *National Union Fire Insurance Co. of Pittsburgh, Pa. v. U.S. Liquids, Inc.*,[96] the Fifth Circuit held that under Texas law the pollution exclusion barred D&O coverage in a shareholder suit. Shareholders alleged that they had acquired U.S. Liquids stock at inflated prices due to the company's concealment from the SEC of its illegal disposition of hazardous wastes. The court found that all claims asserted by the shareholders had one genesis, namely, pollution, and that there was no break in the causal chain linking the underlying damages to that pollution.[97] Thus, even though the underlying claims arose out of deceit regarding pollution, not the pollution itself, the court applied the pollution exclusion. These precedents demonstrate the jurisdiction-specific interpretation of the pollution exclusion in the D&O context. These distinctions could have implications for whether coverage is available for climate change mismanagement claims against directors and officers.

3. Grants of Pollution Coverage

Although pollution is often discussed in the insurance context as something to be excluded from coverage, some insurers offer policies that specifically cover pollution events. Insureds may seek defense and indemnification for climate change-related claims under such policies.

---

92. *Id.* at 1205–06.
93. No. 97-3367, 1998 WL 774109 (6th Cir. Oct. 13, 1998).
94. *Id.* at *1.
95. The asbestos exclusion in *Owens Corning* stated that the following claims are excluded from coverage under the policy:
> [A]ny claim directly or indirectly including but not limited to shareholder derivative suits and/or representative class action suits based upon[,] arising out of or related to: A) Asbestos or any asbestos related injury or damage; or B) Any alleged act, error, omission or duty involving asbestos, its use, exposure, presence, existence, detection, removal, elimination or avoidance; or C) The use, exposure, presence, existence, detection, removal, elimination or avoidance [of] asbestos in any environment, building or structure.

Id. at *2.
96. 88 F. App'x 725 (5th Cir. 2004).
97. *Id.* at 729.

As noted above in Section C.1.b.(3), however, insureds attempting to obtain coverage under policies that specifically cover pollution events may find themselves in the awkward position of arguing that GHGs constitute pollutants for purposes of those policies while at the same time asserting that GHGs do not constitute pollutants under certain environmental statutes or for purposes of CGL and D&O pollution exclusions. Alternatively, a court, in construing a coverage grant broadly and an exclusion narrowly, theoretically could find that a pollution exclusion does not bar coverage in a CGL policy, climate change-related damage and at the same time could hold that an environmental policy provides coverage for such damages.

> If a court broadly interprets an environmental policy to include GHGs and narrowly construes exclusions in D&O or CGL policies to except GHGs, coverage for GHGs may be stacked.

D. Other Potentially Relevant Exclusions

When the pollution exclusion is not determinative of coverage in GHG cases, certain other typical exclusions in insurance policies may impact the analysis of coverage for climate change-related claims. For example, the D&O bodily injury/property damage exclusion; expected or intended injury exclusion; known loss, known injury or damage, loss in progress clauses; known pollution event and pre-existing condition exclusions; injunctive relief exclusions; punitive, treble, exemplary, or multiple damages exclusion; fines, penalties, and taxes exclusion; and intentional/criminal act exclusions may be relevant to evaluating coverage in the climate change context. Even if an exclusion ultimately applies, however, a carrier might still be obligated to provide a defense while coverage issues are being determined. The potential role of each category of exclusion in the climate change context is discussed in turn below.

1. Bodily Injury and Property Damage Exclusion in D&O Policies

> The bodily injury and property damage exclusion in D&O policies may not bar shareholder claims arising out of an underlying climate change-related tort claim alleging bodily injury and/or property damage.

D&O policies often contain an exclusion for claims "for" or "arising out of" bodily injury, mental anguish, emotional distress, sickness, disease, death of any person, or damage to or destruction of any tangible property including loss of use thereof. The purpose of this exclusion is to require that an insured bring claims for bodily injury or property damage under its CGL policy, which specifically provides coverage for both bodily injury and property damage. As such, claims by tort plaintiffs alleging either bodily injury and/or property damage directly against an organization's directors and officers will not be covered.

The bodily injury and property damage exclusion, however, might not bar shareholder claims arising out of a previous underlying climate change-related tort claim alleging bodily injury and/or property damage. For example, if a defendant had to pay tort damages in a climate change-related tort suit, and that damage award had a material financial impact on the defendant, shareholder mismanagement claims might follow. A bodily injury and property damage exclusion might not bar D&O coverage for such a claim. Courts have held that if the exclusion only excludes claims "for" bodily injury or property damage, it does not exclude claims where the damages "arise out of" bodily injury or property damage.[98] Some D&O policies use the "arising out of" language.

2. Expected or Intended Injury Exclusion

The "expected or intended injury" exclusion found in CGL policies could potentially act to bar coverage for some climate change-related claims based on the rationale that the insured must have known or foreseen that its GHG emissions would result in the claimed injury.[99] Plaintiffs in climate change-related litigation, like *Comer* and *Kivalina*, have argued that the defendants knew or should have known about the potential consequences of their GHG emissions, but failed to promptly and adequately mitigate the impact of their emissions. Such allegations appear to directly implicate the "expected or intended injury" exclusion.

The expected or intended injury exclusion generally bars coverage for "[b]odily injury" or "property damage" expected or intended from the standpoint of the insured. This exclusion does not apply to "bodily injury" resulting from the use of reasonable force to protect persons or property.[100]

Courts do not apply the expected or intended exclusion uniformly, and, thus, there is legal uncertainty about its application in the climate change context. For example,

---

98. JOHN H. MATHIAS, MATTHEW M. NEUMEIER & JERRY J. BURGDOERFER, DIRECTORS AND OFFICERS LIABILITY: PREVENTION, INSURANCE, AND INDEMNIFICATION § 8.07 (Law Journal Press 2000).

99. In most jurisdictions, there is significant overlap and interplay between "occurrence" and the "expected and intended injury" exclusion. See Section B, for an in-depth discussion of the concept of "occurrence." Whether coverage is even triggered will depend upon whether the injury producing "event" was "accidental," and thus an "occurrence." The exclusion focuses on the standpoint of the insured.

100. ISO Form CG 00 01 12 07.

some jurisdictions apply a subjective standard, looking at expected or intended injury from the standpoint of the insured, while others apply an objective test.[101] Some jurisdictions require that the injury be intended even though the exclusion reads "expected *or* intended." Courts also have different standards for the extent to which the particular alleged injury must be specifically foreseen and intended.

New York, for instance, focuses on whether resulting damage was specifically intended. The court in *City of Johnstown*, for example, explained:

> It is not enough that an insured was warned that damages might ensue from its actions, or that, once warned, an insured decided to take a calculated risk and proceed as before. Recovery will be barred only if the insured intended the damages, or if it can be said that the damages were, in a broader sense, "intended" by the insured because the insured knew that the damages would flow directly and immediately from its intentional act.[102]

Consequently, in New York, an expected injury is not necessarily excluded because the focus is only on intended injury. Other jurisdictions have concluded that the term "expected" should not be ignored[103] and that "the purpose of adding the phrase 'neither expected nor intended from the standpoint of the insured' was to broaden the class of excluded injuries beyond intentional injuries."[104] Concluding that "expected" must be given independent meaning, a California court found that the exclusion for expected injuries applies to "injuries that the insured subjectively knew or believed to be practically certain to occur even though the insured did not act for the purpose of causing injury."[105]

Given the differing interpretations of this exclusion, it cannot be relied on as a comprehensive measure for limiting or preventing climate change-related exposures.[106] In jurisdictions applying a subjective standard and/or a requirement that the insurer prove that the damages were specifically intended, it may be difficult to prove the applicability of the expected and intended exclusion in the climate change context. Insureds facing a claim in this area are likely to argue that the harm caused

---

101. The standard in the vast majority of jurisdictions is subjective based upon the wording of the exclusion that the injury must be expected or intended "from the standpoint of the insured." At least one court, however, has applied the objective test. *See* City of Carter Lake v. Aetna Cas. & Sur. Co., 604 F.2d 1052 (8th Cir. 1979) (employing the objective test under Iowa law).

102. *See, e.g.,* City of Johnstown v. Bankers Standard Ins. Co., 877 F.2d 1146, 1150 (2d Cir. 1989).

103. *See, e.g.,* Physicians Ins. Co. of Ohio v. Swanson, 569 N.E.2d 906, 911 (Ohio 1991).

104. *See* Armstrong World Indus. v. Aetna Cas. & Sur. Co., 52 Cal. Rptr. 2d 690, 722 (Cal. Ct. App. 1996) (internal quotations and citations omitted).

105. *Id.*

106. Depending on the facts of the particular case, however, it may be worthwhile to litigate the applicability of this exclusion in the climate change context.

by GHG emissions was unknown until recently. And, while such emissions may be intentional, insureds may assert that the alleged consequences of such emissions were not expected or intended.

> Given courts' differing interpretations of the "expected or intended injury" exclusion, this exclusion may not fully limit climate change-related exposures.

3. Known Loss, Known Injury or Damage, Loss in Progress Clauses

> Unlike the expected or intended injury exclusion, the known loss defenses are applicable despite the insured's intentions, as long as the insured knows of the loss at the time policy becomes effective.

In the climate change context, "known loss," "known injury or damage," and/or "loss in progress" language may act to bar coverage for claims alleging harm caused by decades of GHG emissions. The "known injury or damage" limitation is common in the insuring agreement portion of more recent CGL policies.[107] It specifies that the insurer will only cover damages learned by the insured after the effective date of the policy, (i.e., the insurer will not cover any loss which was known to the insured when it contracted for the policy with the insurer). In addition, some liability policies may contain a "loss in progress" exclusion that excludes from coverage *any damage*, known or unknown, which "incepts" before the effective date of the policy.[108] The language essentially codifies the known loss doctrine applied in most jurisdictions and summarized in *Missouri Pacific Railroad Co. v. American Home Assurance Co.*[109]

These known loss defenses are distinct from the expected or intended defense. The known loss defenses are applicable when the insured knows that the loss has already happened or is happening at the time the policy becomes effective. Therefore, it is immaterial whether the loss was accidental, fortuitous, or expected and/or

---

107. The known injury or loss endorsement was written by the ISO in response to the holding in *Montrose Chemical Corp. v Admiral Insurance Co.*, 913 P.2d 878 (Cal. 1995) (discussed below). The language was subsequently added to the CGL policy form.

108. *See, e.g.*, Cummings v. Omaha Prop. and Cas. Ins. Co., 51 F.3d 1043 (5th Cir. 1995) (applying loss-in-progress exclusion to entire claim in which insured bought policy after one building had flooded with intent to insure adjacent building).

109. 675 N.E.2d 1378, 1385 (Ill. App. Ct. 1997) ("[S]uppose an individual, while standing in his basement in three feet of water, calls an insurance company to obtain flood insurance. . . . In such a case, the insurer would not have to show that the insurance policy excluded known losses to avoid liability; it would have to show only that the insured knew that there was a substantial probability that he would suffer a loss from the water.").

intended by the insured at the time of the act or omission that caused the loss. See Sections B and D.2, discussing occurrence and the expected and intended exclusion.

The known loss limitations/exclusions were introduced in response to the decision in *Montrose Chemical Corp. v. Admiral Insurance Co.*, an environmental case, in which the California Supreme Court held that unless explicitly agreed otherwise, (1) a loss is not known to an insured unless at the time the insurer entered into the contract the insured had a legal obligation to pay damages to a third party in connection with the loss and (2) uninterrupted, repeated events may repeatedly trigger coverage if the events are occurrences.[110] Before the development of new endorsements to address erosion of the known loss doctrine in *Montrose* and other cases, the known loss doctrine was primarily a common law doctrine.

In the *AES Corp. v. Steadfast Insurance Co.* litigation, Steadfast denied coverage on three grounds including that the alleged injuries arose before Steadfast's coverage incepted.[111] Steadfast argued that its 2003 through 2008 policies provided no coverage because the *Kivalina* plaintiffs alleged that the damage caused by AES's GHG emissions has been occurring for decades, and those policies all contain a "loss in progress" exclusion.[112] In response, AES argued that the loss in progress exclusion is void in the 2003 through 2008 policies because it conflicts with the trigger limitation and, as such, the provisions are ambiguous and must be interpreted in favor of AES.[113] Specifically, AES argued that the applicable trigger limitation in the policies (excluding coverage for damage *known* to AES before 2003) and the policies' loss in progress exclusion (excluding *all* damage prior to 2003) present an irreconcilable conflict.[114] In addition, AES argued that Steadfast ignored the 1996 through 2000 policies, which do not contain a loss in progress exclusion.[115] According to AES, because the company did not know of the damage its emissions were allegedly causing, any such damage is covered by those policies pursuant to the continuous trigger doctrine.[116] The court did not reach the known loss exclusion issue in *AES* because it decided that there was no occurrence.

As the *AES* litigation illustrates, the effect of the known loss/event trigger limitation and the loss in progress exclusion, along with the interpretation of "occurrence" and the pollution exclusion, may be litigated in the climate change context. A court

---

110. Montrose Chem. Corp. v. Admiral Ins. Co., 913 P.2d 878, 901–05 (Cal. 1995).

111. *See* Chapter 6.F (case overview); Chapter 8.B.1 (occurrence); Chapter 8.C.1.b(2) (definition of "pollutant" and pollution exclusion).

112. *See* Complaint for Declaratory Relief at 6–8, Steadfast Ins. Co. v. AES Corp., No. 2008-858 (Va. Cir. Ct. Arlington County July 9, 2008); Steadfast's Memorandum in Support of Motion Summary Judgment at 13, *Steadfast Ins. Co.*, No. 2008-858 (Va. Cir. Ct. Arlington County Mar. 5, 2009).

113. *See* AES's Opposition to Steadfast's Motion Summary Judgment at 37–38, *Steadfast Ins. Co.*, No. 2008-858 (Va. Cir. Ct. Arlington County Aug. 7, 2009).

114. *See id.*

115. *See id.* at 36.

116. *See id.*

might assess whether a policy has a "loss in progress" exclusion, known loss trigger limitation, or both. Moreover, there may be long-tail claim exposure where the policy only contains the known event/loss trigger limitation. See Sections B.1 and C.2.

> The known loss/event trigger limitation and the loss in progress exclusion may be at issue in climate change coverage litigation.

4. Preexisting Conditions Exclusion

> The preexisting conditions exclusion focuses on the cause of damage rather than the result of the damage and theoretically may bar coverage for some climate change-related claims if the pollution event was known to the insured before the effective date of the policy.

In contrast to the known loss defenses, the "preexisting conditions" exclusion included in some environmental liability policies focuses on the condition (cause of damage) rather than the loss (result).[117] This exclusion bars coverage for pollution events known to the insured before the effective date of the applicable policy. Some policies provide an exception to the exclusion for preexisting conditions disclosed to the insurer on the application for insurance. Accordingly, this exclusion could bar coverage of some climate change-related claims, if there are applicable "pollution events" or "conditions" that were known to the insured before the effective date of the policy.

Assuming that GHG emissions constitute a "pollution event,"[118] the preexisting condition exclusion could potentially bar coverage for some climate change-related claims, given that organizations usually know they are emitting GHGs. Further, some courts have applied this exclusion when the condition at issue existed before the policy incepted, but then became more significant after coverage began.[119]

The preexisting condition exclusion, however, could prove difficult and costly to apply to climate change-related claims. This is true for three reasons. First, some

---

117. *See, e.g.*, Transp. Ins. Co. v. Regency Roofing Cos., Inc., No. 07-80830, 2007 WL 2904156 (S.D. Fla. Oct. 4, 2007) (denying summary judgment on basis of preexisting condition exclusion where insured knew of water leaks before policy incepted, but there was no known resulting mold damage until after the policy incepted).

118. Whether the emission of GHGs can be considered a "pollution event" is discussed in Chapter 8.C.1.b(3), 8.C.2.b, and 8.C.3.

119. *See, e.g.*, Advanced Micro Devices, Inc. v. Great Am. Surplus Lines Ins. Co., 199 Cal. App. 3d 791 (Cal. App. 1988) (where policy included a preexisting conditions exclusion, court ruled losses that are "highly expectable" are not covered). In *Advanced Micro Devices*, before the inception of the policy, the insured was put on notice that its sewer discharge contained levels of chlorinated hydrocarbons that exceeded environmental regulations. The evidence revealed that the plaintiff had actual knowledge of a problem of chlorinated hydrocarbon contamination before the inception of the policy. The only uncertainty, concerned the extent of the problem. Thus, the court found that the plaintiff knew when its insurance began that it faced a costly cleanup project.

courts apply preexisting condition exclusions only if the resulting harm (and not the condition causing the harm) was known to exist before the policy inception.[120] Second, as an exclusionary clause, the insurer has the burden of proof to establish its applicability (in contrast to the Known Injury or Loss clause located in the CGL Insuring Agreement).[121]

Third, it may be difficult to prove that a particular insured versus a particular industry knew about the GHG emissions. For occurrence-based policies it may be especially difficult to prove that the insured knew of the condition at the critical time. For instance, the anti-stacking provision contained in some insurance policies provides that one limit applies to losses occurring over multiple policy periods and that coverage should be provided in the policy period for the date of first exposure to the pollution event that results in a loss. Thus, coverage may be pushed back to a time when the insured did not know about the condition. As a result, insureds may argue that a literal interpretation of this exclusion frustrates the purpose of the policy and provides illusory coverage.[122] Insureds may also assert either that the exclusion does not apply in the climate change context, or that it should be limited to knowledge of preexisting harm, similar to the known loss limitation.

5. Injunctive Relief Exclusion

A policy exclusion for the costs of complying with injunctive or other nonmonetary relief may be relevant to climate change-related claims because plaintiffs may seek not only damages due to the alleged harm caused by GHG emissions, but also injunctive or equitable relief in the form of plant closures or pollution control upgrades. For example, in *AEP v. Connecticut*, the plaintiffs sought only injunctive relief; plaintiffs requested that the court enjoin certain coal-fired power plants from continuing to create a public nuisance and take efforts to abate that nuisance. (See Chapter 6.A.2). If such an exclusion is applicable to these kinds of claims, the costs of complying with equitable relief of this nature may not be covered. The trend recently, however, has been to delete such exclusions from environmental liability policies. Thus, this exclusion may have limited applicability in the climate change context.

---

120. *See, e.g., Transp. Ins. Co.*, 2007 WL 2904156; *see also* Cincotta v. Nat'l Flood Insurers Ass'n, 452 F. Supp. 928 (E.D.N.Y. 1977) (where continuous and repeated condition existed before policy incepted, but damage did not occur until after policy incepted, court found coverage despite exclusion for preexisting conditions).

121. *See, e.g.,* Miller v. Monumental Life Ins. Co., No. CIV 04-0970, 2009 WL 1277745, at *23 (D.N.M. Mar. 31, 2009) ("federal courts have held that the insurer has the burden of proving exclusions, such as for preexisting conditions").

122. *See, e.g.,* Riffe v. Home Finders Assocs., Inc., 205 W. Va. 216, 517 S.E.2d 313, 319 (W. Va. 1999) (reversing summary judgment based on preexisting condition, reasoning that it "would largely nullify the purpose of indemnifying the insured," so its application must be "severely restricted.").

6. Maintenance/Betterment Exclusions

Some environmental liability policies contain exclusions related to maintenance and betterments. Because such exclusions have never been tested in the climate change context, it is unclear whether existing maintenance exclusions are sufficiently comprehensive to address potential climate change-related claims. Given the rise in GHG-related regulations from EPA, insureds may try to seek coverage for installation of GHG monitoring equipment and GHG pollution control technology pursuant to environmental liability policies. Costs of installing GHG monitoring equipment, in particular, may push the limits of the maintenance/betterment exclusion because such equipment may be designed to just measure, not prevent or reduce, GHG emissions. The coverage could turn on whether monitoring constitutes betterment. Emissions reduction measures ordered by courts and other governmental authorities would more plainly fall within such exclusions.

7. Punitive, Treble, Exemplary, or Multiple Damages Exclusion

Although no court has yet awarded damages to a plaintiff for harm caused by climate change, if damages are ever awarded, such damages could include punitive, treble, exemplary, or multiple damages. Exposure to such damages could in part depend on whether the insured's policy contains an exclusion for punitive, treble, exemplary, or multiple damages. Many environmental and professional liability policies often contain such exclusions, and many D&O policies either exclude such damages from the definition of loss or contain an exclusion.

Punitive damages exclusions arose in part because not all states disallow an insurer from insuring against punitive damages. Indeed, there is a split in authority as to whether punitive damages are insurable as a matter of public policy. States that hold punitive measures uninsurable (e.g., Florida, Kansas, and Illinois) do so based on the general rationale that the deterrent effect of the punitive measures would be lost if the cost were shifted to an insurer.[123] States holding punitive measures insurable (e.g., Michigan, Delaware, and Ohio) do so for a variety of reasons, including freedom of contract and protecting compensation for the innocent.[124]

---

123. *See, e.g.*, Int'l Ins. Co. v. Country Manors Ass'n, Inc., 534 So. 2d 1187, 1195–96 (Fla. Dist. Ct. App. 1988) (stating that it is against Florida policy to insure against punitive damages, and therefore, as treble damages awarded against insured are punitive in nature, there is no coverage for such damages in policy).

124. *See, e.g.*, Meijer, Inc. v. Gen. Star Indem. Co., 826 F. Supp. 241, 247 (W.D. Mich. 1993) (if punitive measures were not covered, the policy would provide illusory coverage); Jones v. State Farm Mut. Auto. Ins. Co., 610 A.2d 1352, 1354 (Del. 1992) (punitive measures are insurable to promote policy of compensating the innocent; Corinthian v. Hartford Fire Ins. Co., 758 N.E.2d 218, 222 (Ohio Ct. App. 2001) (allowing coverage for punitive measures promotes freedom of contract).

If an insurer is involved in a coverage action in a state that holds punitive damages insurable, the insurer may be required to cover punitive damages assessed against an insured (assuming the claim is covered and no other exclusions apply) unless it has specifically limited the insuring agreement to non-punitive damages and/or excluded punitive damages from coverage. Further, even in jurisdictions where punitive measures are uninsurable, an insured may argue that the punitive damages at issue are not really punitive in nature and thus are not against public policy. This may occur in instances where punitive damages are assessed based on strict liability or are statutorily required. At least one court has adopted this view, requiring the insurer to cover punitive damages in a state that typically disallows such coverage.[125] Consequently, there may be climate change-related exposure to policies that do not contain a punitive damage exclusion even in jurisdictions that typically disallow coverage of punitive damages.

8. Fines, Penalties, and Taxes Exclusion

Organizations and their officers and directors could in the future face fines or penalties for failure to disclose various risks related to climate change. As EPA promulgates climate change regulations, government-instituted fines, penalties, and even taxes related to GHG emissions and climate change may follow. Already, EPA has set forth a civil penalty structure associated with the recent GHG Reporting Rule. See Chapter 5.B.4. As with punitive damages, exposure in the climate change context could in part depend on whether a policy has a "fines, penalties, and taxes" exclusion. Environmental liability and professional liability policies often contain fines, penalties, and taxes exclusions, and many D&O policies either exclude fines and penalties from the definition of loss or contain a fines and penalties exclusion. Some policies have fines, penalties, and taxes exclusions that do not specifically include punitive damages while other policies have combined fines, penalties, taxes, and punitive damages exclusions.

Regardless of the existence of an exclusion, fines, penalties, and taxes are not always insurable. States that disallow insuring punitive damages usually also disallow insuring fines, penalties, and taxes so long as such measures are punitive in nature.[126] Unlike punitive, treble, exemplary, or multiple damages, however, it is often more difficult to

---

125. *See* Wojciak v. N. Package Corp., 310 N.W.2d 675, 680 (Minn. 1981) (holding statutory punitive damages under workers compensation law insurable because law was enacted not only to punish employers but more importantly as redress to employees who have lost their jobs).

126. State Farm Fire & Cas. Co. v. Martinez, 995 P.2d 890, 895–96 (Kan. Ct. App. 2000) (civil penalties sought by attorney general for lawyer's misconduct are punitive in nature, and thus uninsurable as against public policy); Mortenson v. Nat'l Union Fire Ins. Co. of Pittsburgh, Pa., 249 F.3d 667 (7th Cir. 2001) (holding tax assessment against officer of corporation for failing to remit payroll taxes punitive in nature and thus uninsurable under Illinois's policy against insuring punitive measures).

determine whether a fine, penalty, or tax is indeed punitive.[127] Consequently, unless a policy unambiguously excludes fines, penalties, and taxes, insurers should be prepared for the possibility that they may be required to cover such fines and penalties unless there is another applicable exclusion.

9. Intentional/Criminal Act Exclusions

An intentional/criminal act exclusion is found in most D&O, professional liability, and environmental liability policies. An intentional/criminal act exclusion bars from coverage losses on account of claims made against an insured for fraudulent acts or omissions or willful violations of law. Most such exclusions also contain an adjudication requirement, and some contain a severability clause. The precise wording of the adjudication requirement and severability clause will influence whether this exclusion can reduce potential exposure to climate change-related claims.[128]

There are generally three types of adjudication requirements. The first type states that the exclusion will not apply (i.e., the insurer will be required to defend the insured) until there is an adjudication of wrongdoing. According to this language, the adjudication does not necessarily have to take place in the underlying action but can occur in the context of a declaratory judgment action. Additionally, the adjudication need not be final, in that a judgment at the trial level, before any appeals, will trigger the exclusion. Because the insurer has the right to bring a declaratory judgment action and the exclusion is triggered at the time a trial court renders judgment, this first type of adjudication requirement is the most protective from the perspective of insurers.

The second type of adjudication clause allows an insurer to bring a declaratory judgment action, but requires that there be a final adjudication for the exclusion to be triggered. What constitutes a final adjudication depends on the jurisdiction. One interpretation finds that an adjudication is final when a judgment is had at the trial court—making this type of clause functionally equivalent to clauses requiring only an

---

127. Mitchel v. Cigna Prop. & Cas. Ins. Co., 625 So. 2d 862, 864 (Fla. Dist. Ct. App. 1993) (defining penalty in policy to mean punitive, thus insured's agreement to pay restitution in criminal plea agreement is covered because restitution is not punitive in nature); Littlefield v. Mack, 789 F. Supp. 909, 913 (N.D. Ill. 1992) (attorney fees are covered under policy because not punitive in nature); Carey v. Emp'rs Mut. Cas. Co., 189 F.3d 414 (3d Cir. 1999) (statutory surcharge imposed on township supervisors due to their negligent overpayment to contractor was not punitive in nature and thus not excluded by the policy).

128. In addition, most, if not all states hold as a matter of public policy that criminal acts are uninsurable. Thus, if the intentional act constitutes a crime, the act will not be covered. The public policy against insuring criminal acts is limited, however, in that criminal acts generally cannot be imputed to innocent insureds, limiting the applicability of this policy.

adjudication. Other jurisdictions hold that an adjudication is not final until all appeals have been exhausted.[129]

The third type of adjudication requirement removes the insurer's right to file a declaratory judgment action. Such a requirement will not bar coverage until there is a final adjudication in the underlying action. If the exclusion contains such a requirement, the insurer will be required to defend the insured until there is a judgment in the underlying action, and possibly through the appeals process. For these reasons, insureds prefer this type of adjudication requirement.

Although almost all intentional act exclusions contain an adjudication requirement, some forms of these exclusions may omit the adjudication language. In such circumstances, the insurer may argue that any allegation of intentional acts or fraudulent acts triggers the exclusion. If an insurer is successful in making such an argument, the exclusion would likely preclude potential liability for garden variety climate change-related claims because such claims likely would contain allegations of intentional conduct (i.e., that the insured intentionally emitted GHGs into the atmosphere).

Most intentional/criminal acts exclusion contain a severability clause. A severability clause typically states that the conduct or acts of one insured cannot be imputed to other insureds. Under such a clause the intentional conduct of one insured should not be imputed to another insured who may be liable vicariously, under a strict liability theory, or even negligently.[130] Depending on the language, the acts of one insured also may not be imputed to the insured entity.[131] If an intentional/criminal acts exclusion contains a severability clause, the exclusion will likely have little effect on climate change-related claims because even if one insured's conduct is excluded, it is likely that other insureds' conduct will not be excluded, requiring the insurer to defend and possibly indemnify those insureds. If the exclusion does not contain a severability clause, it is more likely that the exclusion will impact coverage for climate change-related claims.

• •

The intentional/criminal acts exclusion in the climate change context will be limited by the exclusion's final adjudication requirement and the presence of a severability clause.

• •

---

129. *See, e.g., In re* Enron Corp. Secs., Derivative & ERISA Litig., 391 F. Supp. 2d 541, 574 (S.D. Tex. 2005) (implicitly holding that insurer would be required to advance defense costs through appeals); *but see* Pac. Gen. Ins. Co. v. Gen. Dev. Corp., 28 F.3d 1093, 1096 n.7 (11th Cir. 1994) (stating that insurer's obligation to pay defense costs ended when trial court found defendants guilty).

130. *See* J.F. Olson et al., Director and Officer Liability, Indemnification and Insurance § 12:12 (2003).

131. *See id.*

E. Limitations
1. Defense Within or Outside of Limits

> CGL policies generally provide defense outside of limits. Insurers may be exposed to defense costs well above liability limits should they be required to defend climate change-related lawsuits like *Kivalina* and *Comer*.

Whether defense costs are within or outside of policy limits is important with respect to climate change claims because defense costs in actions like *AEP* and *Kivalina* may be high, possibly well above policy limits. The typical CGL policy provides for unlimited defense expenditures that are separate from and in addition to the limits of the policy.[132] In contrast, many professional liability, environmental liability, D&O, and some excess or umbrella policies provide that the limits of indemnity will be reduced by the costs expended in defense of the insured. "Defense within limits policies" may also be called "wasting," "self-consuming," or "self-liquidating" policies, because every dollar spent on defending the claim is one less dollar available to indemnify the insured. Consequently, when "defense is outside of limits," there is a potential for increased exposure to insurers because they might have to pay defense costs on an uncapped basis and, possibly, indemnify the insured up to policy limits.

2. Batch Clauses/Multiple Claim Clauses

"Batch" clauses could be relevant in the climate change-related claim context. Batch clauses, which are also known as related claims or multiple claims clauses, combine, or "batch," all related claims resulting from the same or substantially the same conduct. Batch clauses are often found in professional liability policies, product-liability policies, CGL policies, environmental policies, D&O policies, and umbrella/excess policies.

A typical batch clause provides that any loss (e.g., bodily injury, property damage, etc.) arising from a series of claims as a result of the same or substantially the same occurrence, cause, condition, accident, wrongful act, etc., shall be considered one claim regardless of the number of insureds, claims made, or the number of people or organizations making such claims. Oftentimes the clause requires that all claims shall be deemed to have been made at the time the first of the claims is made. This clause could be found in the conditions or limits of liability section of a policy. In the context of a claims-made policy, this concept also could be dealt with in the definition of claim, related claim, wrongful act, or interrelated wrongful acts.

---

132. ISO Form CG 00 01 12 07.

If an insured is faced with multiple claims relating to its GHG emissions, insurers may be required to indemnify the insured for multiple claims up to the total liability limit if the policy does not have a batch clause limitation. If a batch clause is included in the policy, however, the claims might be considered one claim for purposes of the policy, and insurers might only have to indemnify up to the per claim limit. The individual claim limit is often significantly less than the total liability limit under the policy. Such provisions can also be beneficial to an insured, in that the insured will only need to pay one deductible or retention.

> The inclusion of a batch clause may limit liability for a climate change-related claim to a single limit of liability, regardless of the number of related claims.

3. Anti-Stacking Clauses

"Anti-stacking" clauses, which are related to multiple claims clauses discussed in Section E.2, and are also sometimes called noncumulation clauses,[133] could be important in the climate change-related claim context. Whereas the multiple claim or batch clause provisions discussed in the previous section govern whether multiple occurrences or claims within one policy period are subject to one or more limits of liability and one or more deductible or retention, anti-stacking clauses attempt to govern whether multiple policies can respond to a loss. A typical anti-stacking clause in a CGL policy, for example, states that:

> If this Coverage Form and any other Coverage Form or policy issued to you by us or any company affiliated with us apply to the same "occurrence," the maximum Limit of Insurance under all Coverage Forms or policies shall not exceed the highest applicable Limit of Insurance under any one Coverage Form or policy. This condition does not apply to any Coverage Form or policy issued by us or an affiliated company specifically to apply as excess insurance over this Coverage Form.[134]

The anti-stacking clause above can have two effects depending on the claim. First, if there is a succession of renewal policies, it limits long-tail claims to one policy—that policy with the highest limits of liability. Second, when multiple coverage forms

---

133. The terms batch, noncumulation, anti-stacking, and multiple claims provisions are sometimes referred to interchangeably. For purposes of this discussion, we have separated the discussion of whether a single limit of liability or total limits within *one* policy applies (Section E.2) and whether limits of *multiple* different policies can be stacked (Section E.3).

134. Certain Underwriters at Lloyd's London v. Valiant Ins. Co., 229 P.3d 930, 932 (Wash. Ct. App. 2010).

or policies provide coverage for the same loss (e.g., CGL and professional liability or CGL and environmental), it prohibits recovery under multiple coverages, limiting recovery to the coverage providing the highest limit of liability.[135]

Long-tail claims are common in environmental cases involving damage that is latent, progressive, or deteriorating. An anti-stacking clause may reduce the coverage for climate change-related long-tail claims by limiting coverage to one policy only. However, an anti-stacking provision will not be effective in jurisdictions that apply the continuous trigger theory to such long-tail claims because there will be no, one occurrence.[136] See Section B.3.d. Certain state law trigger and allocation doctrines may trump these clauses.

Depending on the jurisdiction, an anti-stacking provision theoretically could reduce potential liability for climate change-related claims if an insured argues that both its CGL and environmental policies should apply to the claim. Coverage might be limited to one limit of liability only. Conversely, if a policy does not include an anti-stacking provision, insurers may have to pay such claims under multiple policies and multiple limits of liability.[137]

F. General Conditions: Notice Requirement

Notice requirements for occurrence and claims-made and reported policies are similar but different. As discussed in Section A.1, claims-made and reported policies generally require that the insured provide notice of the claim to the insurer during the policy period. The notice requirement in occurrence policies generally require the insured to provide notice of an occurrence "as soon as possible" or "as soon as practicable" and provide notice of an actual claim "immediately." Despite these differences, the notice requirement in both claims-made and occurrence-based policies may have an impact in the context of climate change-related claims. For instance, if an insured becomes aware of a GHG-related claim during the policy period but fails to provide notice of that claim during the policy period or any extended reporting period, an insurer may be able to deny coverage based on such late notice under a claims-made policy. Similarly, under an occurrence policy, an insurer may have a late-notice argument if the

---

135. Although the above example provides that the coverage form or policy with the highest limits of liability will apply, this is not required. An "anti-stacking" clause may specify what policy will apply if multiple policies or coverage forms arguably apply.

136. *See, e.g.,* Spaulding Composites Co. v. Aetna Cas. & Sur. Co., 819, A.2d 410, 420–21 (N.J. 2003) (holding that noncumulation clause in CGL policies inapplicable because under continuous trigger theory, there was no one, single occurrence, but multiple occurrences triggering multiple policies).

137. *See, e.g.,* State v. Cont'l Ins. Co., 88 Cal. Rptr. 3d 288, 309–10 (Ct. App. 2009) *review granted* (holding that in the absence of contract provisions otherwise, stacking is permitted in pollution cases where the damage spans multiple policy periods).

insured waits months or years to report the occurrence to the insurer. As discussed fully in Section B.3, regarding triggers, whether an event constitutes an occurrence is fact, jurisdiction, and possibly court specific. Consequently, it is important to be mindful of the date of the occurrence and when notice of the occurrence is given in relation to that date. It is also important to be mindful that common law standards for when late notice will bar coverage vary from state to state. Further, some jurisdictions have statutory provisions governing notice requirements. Whether a late notice argument will bar coverage in either an occurrence or claims-made policy, therefore, is jurisdiction specific.

G. Will Exposure Drivers Result in Exposure?

The preceding sections reviewed the various policy definitions, exclusions, coverage grants, limitations, and general conditions that are potential exposure drivers in the climate change context. But the pivotal question is, of course, whether any of the enumerated exposure drivers (or any other potential exposure drivers) is likely to result in actual exposure for climate change-related liability to insureds and their insurers.

As discussed in the preceding sections and elsewhere in this book, the answer to this question is likely to depend on a number of key considerations, including:

- *The future path of climate change-related litigation.* Will plaintiffs be able to find a liability theory that sticks, similar to the situation with tobacco, asbestos, and hazardous waste? Or, will issues relating to standing, preemption/displacement, and causation prevent climate change tort litigation from proceeding?
- *Ongoing regulatory activity.* How will EPA's rules regulating GHGs in various contexts fare? Once implemented, will GHG regulations issued by the EPA, state governments, or other entities affect the question of whether GHGs are considered "pollutants" for insurance purposes, or will policyholders succeed in establishing that what qualifies as a "pollutant" in the regulatory context is different from what qualifies as a "pollutant" for insurance purposes?
- *Insurance coverage litigation.* Will the pro-insurer decision in *AES v. Steadfast* dissuade other parties from pursuing insurance coverage for climate change-related claims? Or, might parties pursue coverage litigation in more favorable jurisdictions, potentially leading to more favorable results?
- *The emergence of insurance products tailored to liability for GHG emissions or particular types of renewable energy–related liability. AES v. Steadfast* involved a claim for coverage under the insured's CGL policies. Will new insurance

products emerge that are more specifically tailored to handle climate change-related liability? And, if so, will this lead to a decrease in coverage claims under other types of policies (e.g., CGL, D&O, E&O).
- *The nature and extent of climate change and its effects.* Will the prevailing scientific view that the Earth's atmosphere is warming and stationarity is lost prove correct? Will current predictions concerning a likely increase in the frequency and severity of weather related events linked to global warming be borne out, with a concomitant increase in damage to property and other interests? Will proponents of the theory of anthropogenic climate change establish links between climate change and human activities deemed sufficient for liability/causation purposes?

If past is prologue, the answer to the question of whether climate change is likely to result in significant liability exposure for insureds and their insurers will not emerge overnight, but will take some time to develop. During the time of emergence, of course, potentially affected parties can take actions to influence the ultimate outcomes—actions which could increase or decrease liability. Interested stakeholders, including GHG emitters, "green" product manufacturers and service providers, and their respective insurers, would do well to continue to monitor this area closely and take risk-management actions accordingly.

Glossary of Acronyms & Terms

Term	Definition	Source
AAUs	Assigned Amounts Units; "units issued by Parties to the Kyoto Protocol into their national registry up to their assigned amount, calculated by reference to their base year emissions and their quantified emission limitation and reduction commitment (expressed as a percentage). AAUs are defined in 3/CMP.1, Annex, paragraph 1(b) as follows: An 'assigned amount unit' or 'AAU' is a unit issued pursuant to the relevant provisions in the annex to decision 13/CMP.1 and is equal to one metric tonne of carbon dioxide equivalent, calculated using global warming potentials defined by decision 2/CP.3 or as subsequently revised in accordance with Article 5 (3/CMP.1, Annex, paragraph 1(c))."	*A–Z, Assigned Amount Unit (AAU)*, CDM RULEBOOK, http://cdmrulebook.org/695 (last visited Nov. 17, 2009).
Anthropogenic	"Resulting from or produced by human beings."	IPCC, CLIMATE CHANGE 2007—THE PHYSICAL SCIENCE BASIS 941 (S. Solomon, D. Qin, M. Manning, Z. Chen, M. Marquis, K.B. Averyt, M. Tignor & H.L. Miller eds., 2007), *available at* http://www.ipcc.ch/pdf/assessment-report/ar4/wg1/ar4-wg1-annexes.pdf.

Term	Definition	Source
Atmosphere	"The gaseous envelope surrounding the Earth. The dry atmosphere consists almost entirely of nitrogen (78.1% volume mixing ratio) and oxygen (20.9% volume mixing ratio), together with a number of trace gases, such as argon (0.93% volume mixing ratio), helium and radiatively active greenhouse gases such as carbon dioxide (0.035% volume mixing ratio) and ozone. In addition, the atmosphere contains the greenhouse gas water vapour, whose amounts are highly variable but typically around 1% volume mixing ratio. The atmosphere also contains clouds and aerosols."	IPCC, CLIMATE CHANGE 2007—THE PHYSICAL SCIENCE BASIS 941 (S. Solomon, D. Qin, M. Manning, Z. Chen, M. Marquis, K.B. Averyt, M. Tignor & H.L. Miller eds., 2007), *available at* http://www.ipcc.ch/pdf/assessment-report/ar4/wg1/ar4-wg1-annexes.pdf.
Attribution	The concept of "establishing the most likely causes for . . . detected [climate]change with some defined level of confidence," short of engaging in "controlled experimentation with the climate system."	IPCC, CLIMATE CHANGE 2007—THE PHYSICAL SCIENCE BASIS 668 (Solomon, S., D. Qin, M. Manning, Z. Chen, M. Marquis, K.B. Averyt, M. Tignor & H.L. Miller eds., 2007), *available at* http://www.ipcc.ch/pdf/assessment-report/ar4/wg1/ar4-wg1-chapter9.pdf.
BACT	Best Available Control Technology; "an emissions limitation which is based on the maximum degree of control that can be achieved. It is a case-by-case decision that considers energy, environmental, and economic impact. BACT can be add-on control equipment or modification of the production processes or methods. This includes fuel cleaning or treatment and innovative fuel combustion techniques. BACT may be a design, equipment, work practice, or operational standard if imposition of an emissions standard is infeasible."	*New Source Review Basic Information on PSD*, EPA, http://www.epa.gov/nsr/psd.html#best (last updated July 22, 2011).

Glossary of Acronyms & Terms

Term	Definition	Source
Biomass	"The total mass of living organisms in a given area or volume; dead plant material can be included as dead biomass."	IPCC, CLIMATE CHANGE 2007—THE PHYSICAL SCIENCE BASIS 942 (S. Solomon, D. Qin, M. Manning, Z. Chen, M. Marquis, K.B. Averyt, M. Tignor & H.L. Miller eds., 2007), *available at* http://www.ipcc.ch/pdf/assessment-report/ar4/wg1/ar4-wg1-annexes.pdf.
Black Carbon	"Operationally defined aerosol species based on measurement of light absorption and chemical reactivity and/or thermal stability; consists of soot, charcoal and/or possible light-absorbing refractory organic matter."	IPCC, CLIMATE CHANGE 2007—THE PHYSICAL SCIENCE BASIS 942 (S. Solomon, D. Qin, M. Manning, Z. Chen, M. Marquis, K.B. Averyt, M. Tignor & H.L. Miller eds., 2007), *available at* http://www.ipcc.ch/pdf/assessment-report/ar4/wg1/ar4-wg1-annexes.pdf.
C_2F_6	Hexafluoroethane; an "extremely stable greenhouse gas [...], with [an] atmospheric lifetime estimated at ... 10,000 years.... While [its concentration in the atmosphere] is not large compared to other greenhouse gases ... [its] strong infrared absorption properties and extreme atmospheric inertness make [it] a highly potent greenhouse gas."	Jerry Marks, Ravi Kantamaneni, Diana Pape & Sally Rand, *Protocol for Measurement of Tetrafluoromethane and Hexafluroethane from Primary Aluminum Production* 1, LIGHT METALS 2003 (P. Crepau ed.), *available at* http://www.epa.gov/aluminum-pfc/documents/tms_protocol.pdf.

Glossary of Acronyms & Terms

Term	Definition	Source
CARB	California Air Resources Board; one of six boards, departments, and offices under the umbrella of the California Environmental Protection Agency. "California's Legislature established the Air Resources Board (ARB) in 1967 to: 1. Attain and maintain healthy air quality. 2. Conduct research into the causes of and solutions to air pollution. 3. Systematically attack the serious problems caused by motor vehicles, which are a major cause of air pollution in the State."	*Introduction to the Air Resources Board*, CAL. AIR RES. BD., http://www.arb.ca.gov/html/brochure/arb.htm (last visited Nov. 17, 2009).
Carbon Footprint	The term "carbon footprint" is used to describe the "amount of greenhouse gases that are emitted into the atmosphere each year by a person, household, building, organization, [or] company. . . ." It is usually measured in units of carbon dioxide equivalents.	*What Is a Carbon Footprint?*, EPA, http://climatechange.supportportal.com/link/portal/23002/23006/Article/22042/What-is-a-carbon-footprint-Where-did-this-term-originate (last visited Jan. 27, 2012).
CCS	Carbon Capture and Sequestration; "a broad term that encompasses a number of technologies that can be used to capture carbon dioxide from point sources, such as power plants and other industrial facilities; compress it; transport it mainly by pipeline to suitable locations; and inject it into deep subsurface geological formations for indefinite isolation from the atmosphere."	*CCS Overview: What Is CCS?*, WORLD RES. INST., http://www.wri.org/project/carbon-capture-sequestration/ccs-basics (last visited Nov. 17, 2009).
CDM	Clean Development Mechanism; a market-based mechanism of the Kyoto Protocol, that "allows emission-reduction [or emission-removal] projects in developing countries to earn certified emission reduction (CER) credits, each equivalent to one tonne of CO_2. These CERs can be traded and sold, and used by industrialized countries to a meet a part of their emission reduction targets under the Kyoto Protocol."	*About CDM*, UNFCCC, http://cdm.unfccc.int/about/index.html (last visited Nov. 17, 2009).

Term	Definition	Source
CDP	Carbon Disclosure Project; an independent not-for-profit organization that operates a global climate-change reporting system. Thousands of organizations in some 60 countries around the world now measure and disclose their greenhouse gas emissions and climate change strategies through CDP so they can set reduction targets and make performance improvements. This data is made available by a wide audience of institutional investors, corporations, policymakers and their advisers, public-sector organizations, government bodies, academics, and the public.	Carbon Disclosure Project, https://www.cdproject.net/en-US/Pages/HomePage.aspx (last visited Oct. 1, 2012).
CDSB	Climate Disclosure Standards Board; a consortium of eight business and environmental organizations that works with leading professionals in accountancy, business, standard-setting, and regulation to develop and advocate a generally accepted global framework for use by corporations in disclosing climate change-related information in mainstream reports. CDSB's work program is managed by the Carbon Disclosure Project, which acts as secretariat to CDSB.	*About CDSB*, CLIMATE DISCLOSURE STANDARDS BD., http://www.cdsb.net/about/ (last visited Jan. 27, 2012).
CEMS	Continuous Emissions Monitoring System; a system of "continuous measurement of pollutants emitted into the atmosphere in exhaust gases from combustion or industrial processes. EPA has established requirements for the continuous monitoring of SO_2, volumetric flow, NO_X, diluent gas, and opacity for units regulated under the Acid Rain Program. In addition, procedures for monitoring or estimating carbon dioxide (CO_2) are specified. The CEM rule also contains requirements for equipment performance specifications, certification procedures, and recordkeeping and reporting."	*Continuous Emissions Monitoring Facts Sheet*, EPA, http://www.epa.gov/airmarkt/emissions/continuous-factsheet.html (last visited Nov. 17, 2009).

Term	Definition	Source
CEQ	Council on Environmental Quality; federal agency that "coordinates . . . environmental efforts and works closely with agencies and other White House offices in the development of environmental policies and initiatives. CEQ was established within the Executive Office of the President by Congress as part of the National Environmental Policy Act of 1969 (NEPA) and additional responsibilities were provided by the Environmental Quality Improvement Act of 1970."	*The Council on Environmental Quality—About*, WHITE HOUSE, http://www.whitehouse.gov/administration/eop/ceq/about (last visited Nov. 17, 2009).
CEQA	California Environmental Quality Act; "a statute that requires state and local agencies to identify the significant environmental impacts of their actions and to avoid or mitigate those impacts, if feasible."	*CEQA—Frequently Asked Questions*, CAL. NATURAL RES. AGENCY, http://ceres.ca.gov/ceqa/more/faq.html#what (last visited Nov. 17, 2009).
CER	Certified Emission Reduction; "A Kyoto Protocol unit equal to 1 metric tonne of CO_2 equivalent. CERs are issued for emission reductions from CDM project activities."	*Glossary of Climate Change Acronyms*, UNFCCC, http://unfccc.int/essential_background/glossary/items/3666.php#C (last visited Nov. 17, 2009).
CERCLA	The Comprehensive Environmental Response, Compensation, and Liability Act; "provides a Federal 'Superfund' to clean up uncontrolled or abandoned hazardous-waste sites as well as accidents, spills, and other emergency releases of pollutants and contaminants into the environment. Through CERCLA, EPA was given power to seek out those parties responsible for any release and ensure their cooperation in the cleanup."	*Summary of Superfund*, EPA, http://www.epa.gov/lawsregs/laws/cercla.html (last visited Nov. 17, 2009). *See also* 42 U.S.C. §§ 9601 *et seq.* (2012).
Ceres	Ceres (pronounced "series") is a national network of investors, environmental organizations, and other public-interest groups working with companies and investors to address sustainability challenges like global climate change.	CERES, http://www.ceres.org (last visited Jan. 27, 2012).

Term	Definition	Source
CF_4	Tetrafluoromethane; an "extremely stable greenhouse gas[...], with [an] atmospheric lifetime ... estimated at 50,000 years...." While its concentration in the atmosphere is "not large compared to other greenhouse gases, [its] strong infrared absorption properties and extreme atmospheric inertness make [it] a highly potent greenhouse gas.	Jerry Marks, Ravi Kantamaneni, Diana Pape & Sally Rand, Protocol for Measurement of Tetrafluoromethane and Hexafluoroethane from Primary Aluminum Production 1 (P. Crepeau ed., 2003), *available at* http://www.epa.gov/aluminum-pfc/documents/tms_protocol.pdf.
CFCs	Chlorofluorocarbons; compounds consisting of chlorine, fluorine, and carbon. "CFCs are very stable in the troposphere. They move to the stratosphere and are broken down by strong ultraviolet (UV) light, where they release chlorine atoms that then deplete the ozone layer. CFCs are commonly used as refrigerants, solvents, and foam blowing agents."	*Ozone Layer Protection Glossary*, EPA, http://www.epa.gov/Ozone/defns.html (last visited Nov. 17, 2009).
CH_4	Methane; "a greenhouse gas that remains in the atmosphere for approximately 9–15 years. Methane is over 20 times more effective in trapping heat in the atmosphere than carbon dioxide (CO_2) over a 100-year period and is emitted from a variety of natural and human-influenced sources. Human-influenced sources include landfills, natural gas and petroleum systems, agricultural activities, coal mining, stationary and mobile combustion, wastewater treatment, and certain industrial process[es]."	*Methane*, EPA, http://www.epa.gov/methane/ (last visited Nov. 17, 2009).
Clean Air Act	The law that defines EPA's responsibilities for protecting and improving the nation's air quality and the stratospheric ozone layer.	*Clean Air Act*, EPA, http://www.epa.gov/air/caa (last updated Feb. 17, 2012).

Term	Definition	Source
Climate	"Climate in a narrow sense is usually defined as the average weather, or more rigorously, as the statistical description in terms of the mean and variability of relevant quantities over a period of time ranging from months to thousands or millions of years. The classical period for averaging these variables is 30 years, as defined by the World Meteorological Organization. The relevant quantities are most often surface variables such as temperature, precipitation and wind. Climate in a wider sense is the state, including a statistical description, of the climate system."	IPCC, CLIMATE CHANGE 2007—THE PHYSICAL SCIENCE BASIS 942 (S. Solomon, D. Qin, M. Manning, Z. Chen, M. Marquis, K.B. Averyt, M. Tignor & H.L. Miller eds., 2007), *available at* http://www.ipcc.ch/pdf/assessment-report/ar4/wg1/ar4-wg1-annexes.pdf.
Climate System	"The climate system is the highly complex system consisting of five major components: the atmosphere, the hydrosphere, the cryosphere, the land surface and the biosphere, and the interactions between them."	IPCC, CLIMATE CHANGE 2007—THE PHYSICAL SCIENCE BASIS 943–44 (S. Solomon, D. Qin, M. Manning, Z. Chen, M. Marquis, K.B. Averyt, M. Tignor & H.L. Miller eds., 2007), *available at* http://www.ipcc.ch/pdf/assessment-report/ar4/wg1/ar4-wg1-annexes.pdf.
ClimateWise	Launched in 2007, an initiative composed of insurance companies, whose members commit to the ClimateWise principles: 1. Lead in risk analysis; 2. Inform public policy making; 3. Support climate awareness amongst our customers; 4. Incorporate climate change into our investment strategies; 5. Reduce the environmental impact of our business; and 6. Report and be accountable. Zurich is a member of ClimateWise.	*About*, CLIMATEWISE, http://www.climatewise.org.uk/about/ (last visited Nov. 17, 2009).

Term	Definition	Source
CLRA	Consumer's Legal Remedies Act; prohibits unfair competition and business practices such as "representing that goods or services have sponsorship, approval, characteristics, ingredients, uses, benefits, or quantities which they do not have or that a person has a sponsorship, approval, status, affiliation, or connection which he or she does not have," and allows consumers who suffer from an unfair practice to recover from the violator damages, an injunction, punitive damages, and attorney fees.	Cal. Civ. Code §§ 1770, 1780 (West 2009).
CO_2	Carbon Dioxide; "naturally occurring gas, and also a by-product of burning fossil fuels and biomass, as well as land-use changes and other industrial processes. It is the reference gas against which other greenhouse gases are measured and therefore has a [GWP] of 1."	*Glossary of Climate Change Terms*, EPA, http://www.epa.gov/climatechange/glossary.html#C (last updated June 14, 2012).
CO_2e	Carbon Dioxide Equivalent; a "metric measure used to compare the emissions from various greenhouse gases based upon their [GWP]."	*Glossary of Climate Change Terms*, EPA, http://www.epa.gov/climatechange/glossary.html#C (last updated June 14, 2012).
COP	Conference of the Parties; the supreme body of the United Nations Framework Convention on Climate Change. "It currently meets once a year to review the Convention's progress. The word 'conference' is not used here in the sense of 'meeting' but rather of 'association,'" which explains the seemingly redundant expression "fourth session of the Conference of the Parties."	*Glossary of Climate Change Acronyms*, UNFCCC, http://unfccc.int/essential_background/glossary/items/3666.php#C (last visited Nov. 17, 2009).
CWA	Clean Water Act; a federal law, the objective of which is to "restore and maintain the chemical, physical, and biological integrity of the nation's waters by preventing point and nonpoint pollution sources, providing assistance to publicly owned treatment works for the improvement of wastewater treatment, and maintaining the integrity of wetlands."	*Clean Water Act (CWA)*, EPA, http://www.epa.gov/oecaagct/lcwa.html (last updated June 27, 2012).

Term	Definition	Source
Deforestation	"Conversion of forest to non-forest."	IPCC, Climate Change 2007—The Physical Science Basis 944 (S. Solomon, D. Qin, M. Manning, Z. Chen, M. Marquis, K.B. Averyt, M. Tignor & H.L. Miller eds., 2007), *available at* http://www.ipcc.ch/pdf/assessment-report/ar4/wg1/ar4-wg1-annexes.pdf.
DES	Diethylstilbestrol; a "synthetic form of the hormone estrogen that was prescribed to pregnant women between about 1940 and 1971 because it was thought to prevent miscarriages. Diethylstilbestrol may increase the risk of uterine, ovarian, or breast cancer in women who took it. It also has been linked to an increased risk of clear cell carcinoma of the vagina or cervix in daughters exposed to diethylstilbestrol before birth."	*Dictionary of Cancer Terms*, Nat'l Cancer Inst., http://www.cancer.gov/dictionary/?CdrID=46021 (last visited Nov. 17, 2009).
DJSI	Dow Jones Sustainability Indexes; launched in 1999, the indexes are the first global indexes tracking the financial performance of the leading sustainability-driven companies worldwide.	Dow Jones Sustainability Indexes, http://www.sustainability-index.com/about-us/dow-jones-sustainability-indexes.jsp (last visited Aug. 29, 2012).
DNV	Det Norske Veritas; a global provider of services for managing risk. In particular, DNV was the first entity accredited under the Kyoto Protocol to verify GHG emissions.	Press Release, Det Norske Veritas, *DNV Announces Major Expansion of Climate Change Services in North America*, Reuters, Mar. 30, 2009, *available at* http://www.reuters.com/article/pressRelease/idUS154744+30-Mar-2009+PRN20090330.
DOE	Department of Energy	
EIS	Environmental Impact Statement	

Term	Definition	Source
Emissions Trading	A market-based mechanism of the Kyoto Protocol that "allows countries that have emission units to spare—emissions permitted them but not 'used'—to sell this excess capacity to countries that are over their targets."	*Emissions Trading*, UNFCCC, http://unfccc.int/kyoto _protocol/mechanisms/ emissions_trading/items/2731 .php (last visited Nov. 17, 2009).
EPA	Environmental Protection Agency	
EU ETS	European Union Emissions Trading System; the first international trading system for GHG emissions in the world. It covers more than 11,000 energy-intensive installations across the European Union, representing close to half of Europe's emissions of CO_2. These installations include combustion plants; oil refineries; coke ovens; iron and steel plants; and cement, glass, lime, brick, ceramics, pulp, and paper factories. The aim of the EU ETS is to help member states achieve compliance with their commitments under the Kyoto Protocol.	*European Commission: Emissions Trading System*, EUROPEAN UNION, http:// ec.europa.eu/environment/ climat/emission/index_en.htm (last updated Nov. 15, 2010).
External Forcing	"External forcing refers to a forcing agent outside the climate system causing a change in the climate system."	IPCC, CLIMATE CHANGE 2007—THE PHYSICAL SCIENCE BASIS 945 (S. Solomon, D. Qin, M. Manning, Z. Chen, M. Marquis, K.B. Averyt, M. Tignor & H.L. Miller eds., 2007), *available at* http://www.ipcc.ch/pdf/ assessment-report/ar4/wg1/ ar4-wg1-annexes.pdf.
Extreme Weather Event	"An . . . event that is rare at a particular place and time of year. . . . [A]n extreme weather event would normally be as rare as or rarer than the 10th or 90th percentile of the observed probability density function. By definition, the characteristics of what is called extreme weather may vary from place to place in an absolute sense. . . . When a pattern of extreme weather persists for some time, such as a season, it may be classed as an extreme climate event, especially if it yields an average or total that is itself extreme (e.g., drought or heavy rainfall over a season)."	IPCC, CLIMATE CHANGE 2007—THE PHYSICAL SCIENCE BASIS 945–46 (S. Solomon, D. Qin, M. Manning, Z. Chen, M. Marquis, K.B. Averyt, M. Tignor & H.L. Miller eds., 2007), *available at* http://www.ipcc.ch/pdf/ assessment-report/ar4/wg1/ ar4-wg1-annexes.pdf.

Term	Definition	Source
FTC	Federal Trade Commission	
FTSE	Financial Times Stock Exchange	
FTSE4Good Index Series	The FTSE4Good Index Series has been designed to . . . measure the performance of companies that meet globally recognized corporate responsibility standards" and to facilitate investment in those companies.	*FTSE4Good Index Series*, FTSE, http://www.ftse.com/Indices/FTSE4Good_Index_Series/index.jsp (last visited Nov. 17, 2009).
GAAP	Generally Accepted Accounting Principles	
GHG	"Greenhouse gases are those gaseous constituents of the atmosphere, both natural and anthropogenic, that absorb and emit radiation at specific wavelengths within the spectrum of thermal infrared radiation emitted by the Earth's surface, the atmosphere itself, and by clouds. This property causes the greenhouse effect. Water vapour (H_2O), carbon dioxide (CO_2), nitrous oxide (N_2O), methane (CH_4) and ozone (O_3) are the primary greenhouse gases in the Earth's atmosphere. Moreover, there are a number of entirely human-made greenhouse gases in the atmosphere, such as the halocarbons and other chlorine- and bromine-containing substances, dealt with under the Montreal Protocol. Beside CO_2, N_2O and CH_4, the Kyoto Protocol deals with the greenhouse gases sulphur hexafluoride (SF_6), hydrofluorocarbons (HFCs) and perfluorocarbons (PFCs)."	IPCC, CLIMATE CHANGE 2007—THE PHYSICAL SCIENCE BASIS 947 (Solomon, S., D. Qin, M. Manning, Z. Chen, M. Marquis, K.B. Averyt, M. Tignor & H.L. Miller eds., 2007), *available at* http://www.ipcc.ch/pdf/assessment-report/ar4/wg1/ar4-wg1-annexes.pdf.
Green Building	"[T]he practice of creating structures and using processes that are environmentally responsible and resource-efficient throughout a building's life-cycle from siting to design, construction, operation, maintenance, renovation and deconstruction. This practice expands and complements the classical building design concerns of economy, utility, durability, and comfort. Green building is also known as a sustainable or high performance building."	*Basic Information: Green Building*, EPA, http://www.epa.gov/greenbuilding/pubs/about.htm (last updated Dec. 22, 2010).

Term	Definition	Source
Green Guides	The general principles that apply to all environmental marketing claims and then provide guidance on specific green claims, such as biodegradable, compostable, recyclable, recycled content, and safe for the ozone. The FTC issued the Guides in 1992, and updated them in 1996, 1998, and 2012.	*Reporter Resources: The FTC's Green Guides*, FTC, http://www.ftc.gov/opa/2012/10/greenguides.shtm (last visited Oct. 10, 2012).
Green Marketing	The American Marketing Association defines green marketing as follows: "1. (retailing definition) The marketing of products that are presumed to be environmentally safe. 2. (social marketing definition) The development and marketing of products designed to minimize negative effects on the physical environment or to improve its quality. 3. (environments definition) The efforts by organizations to produce, promote, package, and reclaim products in a manner that is sensitive or responsive to ecological concerns."	*Dictionary*, AM. MKTG. ASS'N, http://www.marketingpower.com/_layouts/Dictionary.aspx?dLetter=G (last visited Nov. 17, 2009).
Greenhouse Effect	"Trapping and build-up of heat in the atmosphere (troposphere) near the Earth's surface. Some of the heat flowing back toward space from the Earth's surface is absorbed by water vapor, carbon dioxide, ozone, and several other gases in the atmosphere and then reradiated back toward the Earth's surface. If the atmospheric concentrations of these greenhouse gases rise, the average temperature of the lower atmosphere will gradually increase."	*Glossary of Climate Change Terms*, EPA, http://www.epa.gov/climatechange/glossary.html#G (last updated June 14, 2012).
Greenwashing	Greenwashing has not been defined by regulation but generally is understood to be "disinformation disseminated by an organisation so as to present an environmentally responsible public image."	CONCISE OXFORD ENGLISH DICTIONARY 624 (10th ed. 2002).

Term	Definition	Source
GWP	Global Warming Potential; "[a]n index, based upon radiative properties of well-mixed greenhouse gases, measuring the radiative forcing of a unit mass of a given well-mixed greenhouse gas in the present-day atmosphere integrated over a chosen time horizon, relative to that of carbon dioxide. The GWP represents the combined effect of the differing times these gases remain in the atmosphere and their relative effectiveness in absorbing outgoing thermal infrared radiation."	IPCC, CLIMATE CHANGE 2007—THE PHYSICAL SCIENCE BASIS 946 (S. Solomon, D. Qin, M. Manning, Z. Chen, M. Marquis, K.B. Averyt, M. Tignor & H.L. Miller eds., 2007), *available at* http://www.ipcc.ch/pdf/assessment-report/ar4/wg1/ar4-wg1-annexes.pdf.
GWSA	Assembly Bill 32—Global Warming Solutions Act, passed by the California Legislature in 2006.	*Assembly Bill 32—Global Warming Solutions Act*, CAL. AIR RES. BD., http://www.arb.ca.gov/cc/ab32/ab32.htm (last visited Nov. 17, 2009).
H_2O	Water Vapor; "[t]he most abundant greenhouse gas, it is the water present in the atmosphere in gaseous form. Water vapor is an important part of the natural greenhouse effect. While humans are not significantly increasing its concentration . . . , it contributes to the enhanced greenhouse effect because the warming influence of greenhouse gases leads to a positive water vapor feedback. In addition to its role as a natural greenhouse gas, water vapor [plays an important role in regulating] the temperature of the planet because clouds form when excess water vapor in the atmosphere condenses to form ice and water droplets and precipitation."	*Glossary of Climate Change Terms*, EPA, http://www.epa.gov/climatechange/glossary.html#W (last updated June 14, 2012).
HCFCs	Hydrochlorofluorocarbons; compounds containing hydrogen, fluorine, chlorine, and carbon atoms. "Although ozone depleting substances, they are less potent at destroying stratospheric ozone than chlorofluorocarbons (CFCs). They have been introduced as temporary replacements for CFCs and are also greenhouse gases."	*Glossary of Climate Change Terms*, EPA, http://www.epa.gov/climatechange/glossary.html#H (last updated June 14, 2012).

Term	Definition	Source
HFCs	Hydrofluorocarbons; compounds containing hydrogen, fluorine, and carbon atoms. "They were introduced as alternatives to ozone depleting substances in serving many industrial, commercial, and personal needs. HFCs are emitted as by-products of industrial processes and are also used in manufacturing. They do not significantly deplete the stratospheric ozone layer, but they are powerful greenhouse gases with global warming potentials ranging from 140 (HFC-152a) to 11,700 (HFC-23)."	*Glossary of Climate Change Terms*, EPA, http://www.epa.gov/climatechange/glossary.html#H (last updated June 14, 2012).
HFEs	Hydrofluorinated Ethers; mainly used in solvents and cleaning. Although HFEs are GHGs, they have a relatively low GWP of 11-14,900. The IPCC has recommended that due to their low GWP, they should be substituted for higher CWP compounds like HFCs.	*Compilation of Technical Information on the New Greenhouse Gases and Groups of Gases Included in the Fourth Assessment Report of the Intergovernmental Panel on Climate Change*, UNFCCC, http://unfccc.int/national_reports/annex_i_ghg_inventories/items/4624.php#Fluorinated (last updated July 27, 2010).
IPCC	Intergovernmental Panel on Climate Change; the leading body for the assessment of climate change, established by the United Nations Environment Programme and the World Meteorological Organization to provide the world with a clear scientific view on the current state of climate change and its potential environmental and socioeconomic consequences.	*Organization*, IPCC, http://www.ipcc.ch/organization/organization.htm (last visited Nov. 17, 2009).

Term	Definition	Source
ISO	International Organization for Standardization; developer and publisher of International Standards. ISO is a network of the national standards institutes of 164 countries, one member per country, with a central secretariat in Geneva, Switzerland, that coordinates the system. ISO is a nongovernmental organization that forms a bridge between the public and private sectors. On the one hand, many of its member institutes are part of the governmental structure of their countries, or are mandated by their government. On the other hand, some members have their roots uniquely in the private sector, having been set up by national partnerships of industry associations.	*About ISO*, ISO, http://www.iso.org/iso/about.htm (last visited November 18, 2009).
JI	Joint Implementation; "defined in Article 6 of the Kyoto Protocol, allows a country with an emission reduction or limitation commitment under the Kyoto Protocol (Annex B Party) to earn emission reduction units (ERUs) from an emission-reduction or emission removal project in another Annex B Party, each equivalent to one tonne of CO_2, which can be counted towards meeting its Kyoto target."	*Joint Implementation*, UNFCCC, http://unfccc.int/kyoto_protocol/mechanisms/joint_implementation/items/1674.php (last visited Nov. 17, 2009).
Kyoto Protocol	"The Kyoto Protocol to the United Nations Framework Convention on Climate Change (UNFCCC) was adopted in 1997 in Kyoto, Japan, at the Third Session of the Conference of the Parties (COP) to the UNFCCC. It contains legally binding commitments, in addition to those included in the UNFCCC. Countries included in the Annex B of the Protocol (most Organisation for Economic Cooperation and Development countries and countries with economies in transition) agreed to reduce their anthropogenic greenhouse gas emissions . . . by at least 5% below 1990 levels in the commitment period 2008 to 2012. The Kyoto Protocol entered into force on February 16, 2005."	IPCC, CLIMATE CHANGE 2007—THE PHYSICAL SCIENCE BASIS 948 (S. Solomon, D. Qin, M. Manning, Z. Chen, M. Marquis, K.B. Averyt, M. Tignor & H.L. Miller eds., 2007), *available at* http://www.ipcc.ch/pdf/assessment-report/ar4/wg1/ar4-wg1-annexes.pdf.

Term	Definition	Source
Land Use	"Land use refers to the total of arrangements, activities and inputs undertaken in a certain land cover type (a set of human actions). The term land use is also used in the sense of the social and economic purposes for which land is managed (e.g., grazing, timber extraction and conservation)."	IPCC, CLIMATE CHANGE 2007—THE PHYSICAL SCIENCE BASIS 948 (S. Solomon, D. Qin, M. Manning, Z. Chen, M. Marquis, K.B. Averyt, M. Tignor & H.L. Miller eds., 2007), *available at* http://www.ipcc.ch/pdf/assessment-report/ar4/wg1/ar4-wg1-annexes.pdf.
LEED	Leadership in Energy and Environmental Design; internationally recognized green building certification system, providing third-party verification that a building or community was designed and built using strategies aimed at improving performance across all the metrics that matter most: energy savings, water efficiency, CO_2 emissions reduction, improved indoor environmental quality, and stewardship of resources and sensitivity to their impacts. Developed by the U.S. Green Building Council (USGBC), LEED provides building owners and operators a concise framework for identifying and implementing practical and measurable green building design, construction, operations and maintenance solutions.	*What LEED Is*, USGBC, http://www.usgbc.org/DisplayPage.aspx?CMSPageID=1988 (last visited Nov. 17, 2009).
MD&A	Management's Discussion and Analysis	
Mitigation	"A human intervention to reduce the sources or enhance the sinks of greenhouse gases."	IPCC, CLIMATE CHANGE 2007—THE PHYSICAL SCIENCE BASIS 949 (S. Solomon, D. Qin, M. Manning, Z. Chen, M. Marquis, K.B. Averyt, M. Tignor & H.L. Miller eds., 2007), *available at* http://www.ipcc.ch/pdf/assessment-report/ar4/wg1/ar4-wg1-annexes.pdf.

Term	Definition	Source
Montreal Protocol	"The Montreal Protocol on Substances that Deplete the Ozone Layer was adopted in Montreal in 1987, and subsequently adjusted and amended in London (1990), Copenhagen (1992), Vienna (1995), Montreal (1997) and Beijing (1999). It controls the consumption and production of chlorine- and bromine-containing chemicals that destroy stratospheric ozone, such as chlorofluorocarbons, methyl chloroform, carbon tetrachloride and many others."	IPCC, CLIMATE CHANGE 2007—THE PHYSICAL SCIENCE BASIS 949 (S. Solomon, D. Qin, M. Manning, Z. Chen, M. Marquis, K.B. Averyt, M. Tignor & H.L. Miller eds., 2007), *available at* http://www.ipcc.ch/pdf/assessment-report/ar4/wg1/ar4-wg1-annexes.pdf.
N_2O	Nitrous Oxide; "powerful greenhouse gas with a global warming potential of 298 times that of carbon dioxide (CO_2). Major sources of nitrous oxide include soil cultivation practices, especially the use of commercial and organic fertilizers, fossil fuel combustion, nitric acid production, and biomass burning."	*Glossary of Climate Change Terms*, EPA, http://www.epa.gov/climatechange/glossary.html#N (last updated June 14, 2012).
NAAQS	National Ambient Air Quality Standards	
NAIC	National Association of Insurance Commissioners	
NEPA	National Environmental Policy Act; "requires federal agencies to integrate environmental values into their decision making processes by considering the environmental impacts of their proposed actions and reasonable alternatives to those actions. To meet NEPA requirements federal agencies prepare a detailed statement known as an Environmental Impact Statement (EIS). EPA reviews and comments on EISs prepared by other federal agencies, maintains a national filing system for all EISs, and assures that its own actions comply with NEPA."	*National Environmental Policy Act: Compliance and Enforcement*, EPA, http://www.epa.gov/Compliance/nepa/ (last visited Nov. 17, 2009).

Term	Definition	Source
NF_3	Nitrogen Trifluoride; a perfluorinated compound that has an estimated atmospheric lifetime of 740 years. It's global warming potential is 17,200 times greater than that of CO_2 when compared over a 100-year period.	IPCC, CLIMATE CHANGE 2007—THE PHYSICAL SCIENCE BASIS 212 (S. Solomon, D. Qin, M. Manning, Z. Chen, M. Marquis, K.B. Averyt, M. Tignor & H.L. Miller eds., 2007), *available at* http://www.ipcc.ch/pdf/assessment-report/ar4/wg1/ar4-wg1-chapter2.pdf
NGO	Nongovernmental Organization	
NHTSA	National Highway Traffic Safety Administration	
NOAA	National Oceanic and Atmospheric Administration	
NRC	National Research Council	
NRD	Natural Resource Damages; "for injury to, destruction of, or loss of natural resources, including the reasonable costs of a damage assessment. Both CERCLA and OPA [Oil Pollution Act] define 'natural resources' broadly to include 'land, fish, wildlife, biota, air, water, ground water, drinking water supplies, and other such resources....' Both statutes limit 'natural resources' to those resources held in trust for the public, termed Trust Resources. While there are slight variations in their definitions, both CERCLA and OPA state that a 'natural resource' is a resource 'belonging to, managed by, held in trust by, appertaining to, or otherwise controlled by' the United States, any State, an Indian Tribe, a local government, or a foreign government."	*Natural Resource Damages: A Primer*, EPA, http://www.epa.gov/superfund/programs/nrd/primer.htm (last updated Aug. 9, 2011). 42 U.S.C. § 9601(16). 33 U.S.C. § 2701(20).
NRDC	Natural Resources Defense Council; an environmentalist NGO with the stated purpose "to safeguard the Earth: its people, its plants and animals and the natural systems on which all life depends."	*NRDC: Mission Statement*, NATURAL RES. DEF. COUNCIL, http://www.nrdc.org/about/mission.asp (last visited Nov. 17, 2009).

Term	Definition	Source
NSR	New Source Review; "NSR requires stationary sources of air pollution to get permits before they start construction. NSR is also referred to as construction permitting or preconstruction permitting."	*New Source Review Basic Information*, EPA, http://www.epa.gov/NSR/info.html (last updated July 22, 2011).
OPR	Office of Planning and Research; provides legislative and policy research support for the California governor's office. OPR also assists the governor and the administration in land-use planning and manages the Office of the Small Business Advocate.	*Governor's Office of Planning and Research: Welcome*, STATE OF CAL., http://www.opr.ca.gov/index.php?a=about/about.html (last visited Nov. 17, 2009).
Ozone	"[T]he triatomic form of oxygen (O_3), is a gaseous atmospheric constituent. In the troposphere, it is created both naturally and by photochemical reactions involving gases resulting from human activities (smog). Tropospheric ozone acts as a greenhouse gas. In the stratosphere, it is created by the interaction between solar ultraviolet radiation and molecular oxygen (O_2). Stratospheric ozone plays a dominant role in the stratospheric radiative balance. Its concentration is highest in the ozone layer."	IPCC, CLIMATE CHANGE 2007—THE PHYSICAL SCIENCE BASIS 950 (S. Solomon, D. Qin, M. Manning, Z. Chen, M. Marquis, K.B. Averyt, M. Tignor & H.L. Miller eds., 2007), *available at* http://www.ipcc.ch/pdf/assessment-report/ar4/wg1/ar4-wg1-annexes.pdf.
Ozone Layer	"The stratosphere contains a layer in which the concentration of ozone is greatest, the so-called ozone layer. The layer extends from about 12 to 40 km above the Earth's surface."	IPCC, CLIMATE CHANGE 2007—THE PHYSICAL SCIENCE BASIS 950 (S. Solomon, D. Qin, M. Manning, Z. Chen, M. Marquis, K.B. Averyt, M. Tignor & H.L. Miller eds., 2007), *available at* http://www.ipcc.ch/pdf/assessment-report/ar4/wg1/ar4-wg1-annexes.pdf.
PAYD	Pay-As-You-Drive	

Term	Definition	Source
PFCs	Perfluorocarbons; "[human-made] chemicals composed of carbon and fluorine only. These chemicals (predominantly CF_4 and C_2F_6) were introduced as alternatives, along with hydrofluorocarbons, to the ozone depleting substances. In addition, PFCs are emitted as by-products of industrial processes and are also used in manufacturing. PFCs do not harm the stratospheric ozone layer, but they are powerful greenhouse gases."	*Glossary of Climate Change Terms*, EPA, http://www.epa.gov/climatechange/glossary.html#P (last updated June 14, 2012).
Positive Feedback Loop	"As the temperature of the atmosphere rises, more water is evaporated from ground storage (rivers, oceans, reservoirs, soil). Because the air is warmer, the relative humidity can be higher (in essence, the air is able to 'hold' more water when its warmer), leading to more water vapor in the atmosphere. As a greenhouse gas, the higher concentration of water vapor is then able to absorb more thermal . . . energy radiated from the Earth, thus further warming the atmosphere. The warmer atmosphere can then hold more water vapor and so on and so on. . . . [There is] scientific uncertainty . . . in defining the extent and importance of this feedback loop."	*Greenhouse Gases: Frequently Asked Questions*, NAT'L OCEANIC & ATMOSPHERIC ADMIN., NAT'L CLIMATIC DATA CTR., http://lwf.ncdc.noaa.gov/oa/climate/gases.html (last updated Feb. 23, 2010).
PSD	Prevention of Significant Deterioration; EPA permitting requirements that apply to "new major sources or major modifications at existing sources for pollutants where the area the source is located is in attainment or unclassifiable with the National Ambient Air Quality Standards (NAAQS). It requires the following: 1. installation of the "Best Available Control Technology (BACT)"; 2. an air quality analysis; 3. an additional impacts analysis; and 4. public involvement."	*New Source Review Basic Information on PSD*, EPA, http://www.epa.gov/nsr/psd.html (last updated July 22, 2011).

Term	Definition	Source
Radiative Forcing	Radiative forcing is the change in the net vertical irradiance (expressed in watts per square meter, W/m^2) at the tropopause due to an internal change or a change in the external forcing of the climate system, such as, for example, a change in the concentration of carbon dioxide or the output of the sun. Usually radiative forcing is computed after allowing for stratospheric temperatures to readjust to radiative equilibrium, but with all tropospheric properties held fixed at their unperturbed values. Radiative forcing is called instantaneous if no change in stratospheric temperature is accounted for.	*Glossary of Climate Change Terms*, EPA, http://www.epa.gov/climatechange/glossary.html#R (last updated June 14, 2012). IPCC, CLIMATE CHANGE 2007—THE PHYSICAL SCIENCE BASIS 951 (Solomon, S., D. Qin, M. Manning, Z. Chen, M. Marquis, K.B. Averyt, M. Tignor & H.L. Miller eds., 2007), *available at* http://www.ipcc.ch/pdf/assessment-report/ar4/wg1/ar4-wg1-annexes.pdf.
RCRA	Resource Conservation and Recovery Act; "gives EPA the authority to control hazardous waste from the 'cradle-to-grave.' This includes the generation, transportation, treatment, storage, and disposal of hazardous waste. RCRA also set forth a framework for the management of non-hazardous solid wastes. The 1986 amendments to RCRA enabled EPA to address environmental problems that could result from underground tanks storing petroleum and other hazardous substances."	*Summary of the RCRA*, EPA, http://www.epa.gov/lawsregs/laws/rcra.html (last updated Aug. 23, 2012). 42 U.S.C. §§ 6901 *et seq.*
RGGI	Regional Greenhouse Gas Initiative; the first mandatory, market-based effort in the United States to reduce greenhouse gas emissions. Ten Northeastern and Mid-Atlantic states have capped and will reduce CO_2 emissions from the power sector 10 percent by 2018.	*Regional Greenhouse Gas Initiative (RGGI) CO_2 Budget Trading Program*, RGGI, http://www.rggi.org/home (last visited Nov. 17, 2009).

Term	Definition	Source
SDWA	Safe Drinking Water Act; "established to protect the quality of drinking water in the U.S. This law focuses on all waters actually or potentially designed for drinking use, whether from above ground or underground sources. The Act authorizes EPA to establish minimum standards to protect tap water and requires all owners or operators of public water systems to comply with these primary (health-related) standards. The 1996 amendments to SDWA require that EPA consider a detailed risk and cost assessment, and best available peer-reviewed science, when developing these standards. State governments, which can be approved to implement these rules for EPA, also encourage attainment of secondary standards (nuisance-related). Under the Act, EPA also establishes minimum standards for state programs to protect underground sources of drinking water from endangerment by underground injection of fluids."	*Summary of the SDWA*, EPA, http://www.epa.gov/lawsregs/laws/sdwa.html (last updated Aug. 23, 2012). 42 U.S.C. § 300f *et seq.*
Sea Ice	"Any form of ice found at sea that has originated from the freezing of seawater. Sea ice may be discontinuous pieces (ice floes) moved on the ocean surface by wind and currents (pack ice), or a motionless sheet attached to the coast (land-fast ice)."	IPCC, CLIMATE CHANGE 2007—THE PHYSICAL SCIENCE BASIS 951 (S. Solomon, D. Qin, M. Manning, Z. Chen, M. Marquis, K.B. Averyt, M. Tignor & H.L. Miller eds., 2007), *available at* http://www.ipcc.ch/pdf/assessment-report/ar4/wg1/ar4-wg1-annexes.pdf.
SEC	Securities and Exchange Commission	
SEPA	State Environmental Policy Act	
SEQA	State Environmental Quality Act	

Term	Definition	Source
Sequestration	"The uptake [addition of a substance of concern to a reservoir] of carbon containing substances, in particular carbon dioxide, is often called (carbon) sequestration."	IPCC, CLIMATE CHANGE 2007—THE PHYSICAL SCIENCE BASIS 952, 954 (S. Solomon, D. Qin, M. Manning, Z. Chen, M. Marquis, K.B. Averyt, M. Tignor & H.L. Miller eds., 2007), *available at* http://www.ipcc.ch/pdf/assessment-report/ar4/wg1/ar4-wg1-annexes.pdf.
SF_6	Sulfur Hexafluoride; "colorless gas soluble in alcohol and ether, slightly soluble in water. A very powerful greenhouse gas used primarily in electrical transmission and distribution systems and as a dielectric in electronics. The global warming potential of SF_6 is 22,800."	*Glossary of Climate Change Terms*, EPA, http://www.epa.gov/climatechange/glossary.html#S (last updated June 14, 2012).
Sink	"Any process, activity or mechanism that removes a greenhouse gas, an aerosol or a precursor of a greenhouse gas or aerosol from the atmosphere."	IPCC, CLIMATE CHANGE 2007—THE PHYSICAL SCIENCE BASIS 952 (S. Solomon, D. Qin, M. Manning, Z. Chen, M. Marquis, K.B. Averyt, M. Tignor & H.L. Miller eds., 2007), *available at* http://www.ipcc.ch/pdf/assessment-report/ar4/wg1/ar4-wg1-annexes.pdf.
Solar Radiation	Radiation emitted by the sun. It is also referred to as short-wave radiation. Its distinctive range of wavelengths (spectrum) is determined by the sun's temperature.	*Glossary of Climate Change Terms*, EPA, http://www.epa.gov/climatechange/glossary.html#S (last updated June 14, 2012).
Soot	"Particles formed during the quenching of gases at the outer edge of flames of organic vapors, consisting predominantly of carbon, with lesser amounts of oxygen and hydrogen present as carboxyl and phenolic groups and exhibiting an imperfect graphitic structure."	IPCC, CLIMATE CHANGE 2007—THE PHYSICAL SCIENCE BASIS 952 (S. Solomon, D. Qin, M. Manning, Z. Chen, M. Marquis, K.B. Averyt, M. Tignor & H.L. Miller eds., 2007), *available at* http://www.ipcc.ch/pdf/assessment-report/ar4/wg1/ar4-wg1-annexes.pdf.

Term	Definition	Source
Stationarity	The idea that natural systems fluctuate within an unchanging envelope of variability.	P.C.D. Milly et al., *Stationarity Is Dead: Whither Water Management?*, 319 SCIENCE 573–74 (2008).
Superfund	"Superfund is the name given to the environmental program established to address abandoned hazardous waste sites. It is also the name of the fund established by the Comprehensive Environmental Response, Compensation and Liability Act of 1980, as amended. This law was enacted in the wake of the discovery of toxic waste dumps such as Love Canal and Times Beach in the 1970s. It allows the EPA to clean up such sites and to compel responsible parties to perform cleanups or reimburse the government for EPA-[led] cleanups."	*Superfund: Basic Information*, EPA, http://www.epa.gov/superfund/about.htm (last updated May 14, 2012).
UNEP	United Nations Environment Programme; agency whose stated mission is "to provide leadership and encourage partnership in caring for the environment by inspiring, informing, and enabling nations and peoples to improve their quality of life without compromising that of future generations."	*About UNEP: The Organization*, UNEP, http://www.unep.org/Documents.Multilingual/Default.asp?DocumentID=43 (last visited Nov. 17, 2009).

Term	Definition	Source
UNFCCC	United Nations Framework Convention on Climate Change; "sets an overall framework for intergovernmental efforts to tackle the challenge posed by climate change. It recognizes that the climate system is a shared resource whose stability can be affected by industrial and other emissions of carbon dioxide and other greenhouse gases. The Convention enjoys near universal membership, with 192 countries having ratified. Under the Convention, governments: gather and share information on greenhouse gas emissions, national policies and best practices; launch national strategies for addressing greenhouse gas emissions and adapting to expected impacts, including the provision of financial and technological support to developing countries; cooperate in preparing for adaptation to the impacts of climate change. The Convention entered into force on March 21, 1994."	*The United Nations Framework Convention on Climate Change*, UNFCCC, http://unfccc.int/essential_background/convention/items/2627.php (last visited Nov. 17, 2009).
USGCRP	The U.S. Global Change Research Program; "coordinates and integrates federal research on changes in the global environment and their implications for society. The USGCRP began as a presidential initiative in 1989 and was mandated by Congress in the Global Change Research Act of 1990 . . . which called for 'a comprehensive and integrated United States research program which will assist the Nation and the world to understand, assess, predict, and respond to human-induced and natural processes of global change.'"	*Program Overview*, U.S. GLOBAL CHANGE RESEARCH PROGRAM, http://www.globalchange.gov/about (last visited July 16, 2012).

Term	Definition	Source
WBCSD	World Business Council for Sustainable Development; a CEO-led, global association of some 200 companies dealing exclusively with business and sustainable development. It provides a platform for companies to explore sustainable development; share knowledge, experiences and best practices; and to advocate business positions on these issues in a variety of forums, working with governments, nongovernmental and intergovernmental organizations. Members are drawn from more than 35 countries and 20 major industrial sectors. The council also benefits from a global network of some 60 national and regional business councils and regional partners.	WBCSD, http://wbcsd.org/about.aspx (last visited Jan. 27, 2012).
WCI	Western Climate Initiative; "a collaboration of independent jurisdictions who commit to work together to identify, evaluate, and implement . . . policies to tackle climate change at a regional level." Members include Arizona, British Columbia, California, Manitoba, Montana, New Mexico, Ontario, Oregon, Quebec, Utah, and Washington.	*Organization*, WCI, http://www.westernclimateinitiative.org/organization (last visited Nov. 17, 2009).
WMO	World Meteorological Organization; "a specialized agency of the United Nations. It is the UN system's authoritative voice on the state and behavior of the Earth's atmosphere, its interaction with the oceans, the climate it produces and the resulting distribution of water resources."	*WMO in Brief*, WMO, http://www.wmo.int/pages/about/index_en.html (last visited Nov. 17, 2009).
WRI	World Resources Institute; an environmental think tank with the stated mission "to move human society to live in ways that protect Earth's environment and its capacity to provide for the needs and aspirations of current and future generations."	*Who We Are*, WRI, http://www.wri.org/about (last visited Nov. 17, 2009).

Index

A

Absolute pollution exclusion, 172–175
Accident coverage, 8
Acid Rain Program, 46
Adaptation, 1
Advertising, 123–128, 142
AES Corp. v. Steadfast Insurance Co., 6, 92–93, 142, 156–158, 168–169
AES Corporation, 107
Air pollutants, 40
Air pollution, 41
Ajax, 126
Alito, Samuel, 74
Alternative energy, 53–56
American Clean Energy and Security Act, 51
American Electric Power Co., Inc. v. Connecticut, 3, 23n7
    defense costs in, 191
    environmental liability coverage and, 143
    federal regulation of GHGs and, 51
    as first climate change related tort litigation, 70
    future injuries associated with climate change cited in, 26
    Ninth Circuit decision in *Native Village of Kivalina v. ExxonMobil Corporation* and, 79
    overview of, 71–76
    pollution exclusions and, 176
    suit seeking to curtail GHG emissions, 25
American Home Products Corp. v. Liberty Mutual Insurance Co., 162
American Recovery and Reinvestment Act (ARRA), 54
Anhydrous ammonia, 165
Annual reports, 115
Anti-stacking clauses, 192–193
Arizona Law Review, 22
ARRA. *See* American Recovery and Reinvestment Act
Article III standing, 73, 74, 75, 77, 78, 80
Asbestos litigation, 3, 22, 90
Assembly Bill 1493, 57
Assembly Bill 32, 59
Attribution
    as defined by the IPCC, 15–16
    research, 13

B

BACT. *See* Best achievable control technology
Bad faith claims, 146–147
Bali Action Plan, 36
Batch clauses, 191–192
Berlin Mandate, 34
Bernstein v. Toyota Motor Sales USA, Inc., 125
Best achievable control technology (BACT), 47, 49, 50, 55
Betterment exclusions, 143, 187
Black carbon, 17, 41
Bodily injury and property damage exclusion, 140–141
Boiler and machinery insurance, 150
Boxer, Barbara, 51
Builders-risk policies, 137–138, 150
Bush administration, 38
Business interruption coverage, 137, 150
Buzz Off Insect Shield, LLC, 127

C

CAFE standards. *See* Corporate average fuel economy standards
California
    Assembly Bill 1493, 57
    Assembly Bill 32, 59
    California Air Resources Board, 57, 58, 119
    California Environmental Quality Act, 59, 117–119

California (*continued*)
   California Green Insurance Act, 66–67
   California v. General Motors Corp., 3, 80–81
   cap-and-trade regulation, 59, 118, 120
   Consumer Legal Remedies Act, 125, 126
   False Advertising Law, 126
   Global Warming Solutions Act, 58
   Insurer Climate Change Disclosure Survey, 66
   legislation regulating GHG limits for vehicles, 57–59
   Natural Resources Agency, 118
   Office of Administrative Law, 118
   Office of Planning and Research, 118
   Pay-As-You-Drive regulations, 67
   Senate Bill 97, 118
   Unfair Competition Law, 125, 126, 128
   use of market share liability theory, 89
California Air Resources Board (CARB), 57, 58, 119
California Climate Action Registry, 114
California Court of Appeal, Fouth District, 124
California Environmental Quality Act (CEQA), 57, 59, 117–119
California Green Insurance Act, 66–67
California Supreme Court, 89
California v. General Motors Corp., 3, 80–81, 143
Cancun Agreements, 37
Cap-and-trade regulation
   California, 59, 118
   disputes in carbon and GHG markets, 120–122
   Quebec, 120
   Regional Greenhouse Gas Initiative, 60–61
CARB. *See* California Air Resources Board
Carbon capture and sequestration (CCS)
   new fossil-fuel-fired power plants and, 51
   policies, 8
   pollution exclusion for carbon dioxide, 169
   regulation and potential litigation related to, 52–53
Carbon dioxide (CO_2)
   in *AES Corp. v. Steadfast Insurance Co.*, 169
   cap-and-trade regulation for, 60–61
   carbon capture and sequestration, 52–53
   Cause or Contribute Finding for, 167
   Clean Air Act emission thresholds for, 168
   Climate Registry reporting protocol for, 114
   in definition of air pollutant, 40
   in definition of air pollution, 41
   emission reduction credits for, 35
   Endangerment Finding for, 167
   GHG emissions reports for, 26, 45, 167
   Kivalina v. ExxonMobil Corp., 76
   light-duty vehicle emission standards for, 167, 168
   percentage of U.S. emissions, 17
   pollution exclusion for, 169
   regulation under Tailoring Rule, PSD and Title V programs, 47
   as source targeted for regulation, 16, 17–18
Carbon dioxide equivalent. *See* CO_2 equivalent
Carbon Disclosure Project (CDP), 2, 108–110
Carbon markets
   disputes arising out of, 120–122
   European Union Emission Trading System, 120–122
   insurers and, 36
   Kyoto Protocol and, 35
Carbon monoxide, 165
Carbon storage. *See* Carbon capture and sequestration
Causation issue
   attribution, 15–16, 88
   greenhouse gases and, 14–16
   market share liability theory and, 88–90
   NRD claims and, 96
Cause or Contribute Finding, 40–41
CCS. *See* Carbon capture and sequestration
CDM. *See* Clean development mechanism
CDSB. *See* Climate Disclosure Standards Board
CEMS. *See* Continuous emissions monitoring system
Center for Climate Change Law, 91
CEQ. *See* Council on Environmental Quality
CEQA. *See* California Environmental Quality Act
CERCLA. *See* Comprehensive Environmental Response, Compensation, and Liability Act
Ceres, 28, 98, 99, 102
CERs. *See* Certified emissions reduction credits
Certified emissions reduction credits (CERs), 35
CFCs. *See* Chlorofluorocarbons
CGL policies. *See* Commercial general liability policies

CH$_4$. *See* Methane
China
    Durban Platform for Enhanced Action and, 38
    Kyoto Protocol and, 34, 36
Chlorofluorocarbons (CFCs), 19, 41
Chrysler Group LLC, 80
Cinergy Corp. v. Associated Electric & Gas Insurance Services, Ltd., 159–160
Circumstance reporting, 6–7, 143
City of Albuquerque, 131
Claims
    arising out of corporate disclosure and management of climate change risk, 97–115
    design-defect, 142
    failure to warn, 142
    federal common-law nuisance, 51, 71, 75, 77, 79
    greenwashing litigation, 122–129
    long-tail, 164
    for natural resource damages pursuant to CERCLA, 95–97
    potential exposure drivers, 153–195
    products liability, 87–88, 142
Claims-made policies
    conditions for coverage, 154–155
    environmental liability coverage and, 6
    extended reporting period coverage, 155
    policy period coverage, 155
    retroactive date requirement, 156
Clean Air Act
    American Electric Power Co., Inc. v. Connecticut, 73–75
    California request for waiver, 58
    environmental liability coverage and, 143
    EPA action pursuant to, 38–43, 167, 168
    meaning of the term "pollutant" in, 168
    renewable energy and, 55
    Section 202(a)(1), 39–40, 42
    Title V, 44, 47
Clean development mechanism (CDM), 35
Clean Energy Jobs & American Power Act, 51
Clean Water Act (CWA), 78
Clifton, Richard, 80

Climate change
    circumstance reporting of events, 143
    claims arising out of management of risk, 97–115
    definition in the insurance context, 13–14
    as an emerging risk, 21
    potential litigation arising out of efforts to address, 115–134
    shareholder resolutions related to, 98–102
    use of National Environmental Policy Act to address, 115–117
    use of state environmental policy acts to address, 117–120
Climate Disclosure Standards Board (CDSB), 109–110
Climate Registry, 114–115
Climate Risk Disclosure by Insurers: Evaluating Insurer Responses to the NAIC Climate Disclosure Survey, 63
ClimateWise, 148
CLRA. *See* Consumer Legal Remedies Act
CO$_2$. *See* Carbon dioxide
CO$_2$ equivalent (CO$_2$e), 28, 47, 97
Coalition for Responsible Regulation v. EPA, 42, 44, 50
Colgate-Palmolive, Inc., 126
Colony National Insurance Co. v. Specialty Trailer Leasing, Inc., 166
Columbia Law School, 91
Comer, Ned, 86
Comer v. Murphy Oil USA, Inc., 3, 22, 25, 82–87, 162
Commercial general liability (CGL) policies
    in advertising and professional contexts, 132–134
    anti-stacking clause in, 192–193
    batch clauses in, 191
    claims-made and reported trigger in, 154
    continuous trigger theory and, 163–164
    defense within or outside of limits in, 191
    effect of the trigger on occurrence in, 160
    greenwashing claims and, 123, 127–128
    implications for, 5, 6, 141–142
    issues in disputes, 142
    for renewable energy projects, 151
    retroactive date requirement, 156

Comprehensive Environmental Response,
    Compensation, and Liability Act
    (CERCLA)
  claims for natural resource damages pursuant
    to, 95–97
  commercial general liability policies and, 142
  environmental liability coverage and, 143
  implications for environmental liability
    coverage and, 6
  litigation, 22
  pollution exclusion for carbon dioxide and, 169
Conference of the Parties (COP)
  Bali Action Plan, 36
  Berlin Mandate, 34
  Cancun Agreements, 37
  Copenhagen Accord, 37
  Durban Platform for Enhanced Action, 37–38
  Poznán, Poland, 37
Consumer Legal Remedies Act (CLRA),
    125, 126
Contingent business interruption coverage, 137
Continuous emissions monitoring system
    (CEMS), 46
Continuous trigger theory, 163–164, 193
Contractors protective professional indemnity
    and liability insurance (CPPI), 145–146
COP. *See* Conference of the Parties
Copenhagen Climate Change Conference, 37
Corporate average fuel economy (CAFE)
    standards, 43
Corporate reimbursement coverage. *See* Side B
    coverage
Council on Environmental Quality (CEQ),
    116–117
Court of Appeals of Indiana, 159
CPPI. *See* Contractors protective professional
    indemnity and liability insurance
Cuomo, Andrew, 107
CWA. *See* Clean Water Act

D

D&O coverage. *See* Directors and officers
    coverage
Delay in completion coverage, 137–138
Department of Environmental Conservation, 126

DES. *See* Diethylstilbestrol
Design-defect claims, 142
Det Norske Veritas (DNV), 121–122
Detection, 15n22
Diethylstilbestrol (DES), 89–90
Directors and officers coverage (D&O coverage)
  actions by regulators, 138–139
  actions by shareholders, 140
  bodily injury/property damage exclusion,
    180–181
  claims-made and reported trigger in, 154
  disclosure irregularities and, 97
  disclosure liability and, 102
  extended reporting period for, 155
  greenwashing claims and, 123
  implications for, 138–146
  policy period for, 155
  potentially relevant provisions, 140–141
  retroactive date requirement, 156
  special issues associated with pollution
    exclusions in, 178–179
Disclosure
  annual reports, 115
  Carbon Disclosure Project, 108–110
  claims arising out of corporate, 97–98
  Climate Registry, 114–115
  Dow Jones Sustainability Indexes,
    111–112
  FTSE4Good Index Series, 112–113
  liability based on, 102, 147–148
  obligations imposed by Securities Act of
    1933, 104
  obligations imposed by Securities Exchange
    Act of 1934, 104
  Principles for Responsible Investment,
    110–111
  role of SEC in climate risk, 102
  securities laws, regulations, and guidance
    requiring, 104–106
  shareholder resolutions, 98–102
  voluntary programs, 108–115
DJSIs. *See* Dow Jones Sustainability Indexes
DNV. *See* Det Norske Veritas
DOE. *See* U.S. Department of Energy
Dominion Resources, Inc., 107

Dow Jones Sustainability Indexes (DJSIs), 2, 111–112
Durban Platform for Enhanced Action, 37–38
Dyna-E International, Inc., 128
Dynegy, Inc., 107

E

E&O coverage. *See* Professional liability coverage
EGUs. *See* Electric utility generating units
Electric utility generating units (EGUs), 50–51, 55
Electricity generation, 53
Emerging risks, climate change as, 21
Emissions-reduction units (ERUs), 35
Emissions trading, 35
Endangerment Finding, 40–43
Energy Policy Act, 54
Energy Policy and Conservation Act, 43, 124
Energy Tax Prevention Act, 38
Enforcement actions, 138–139
Enterprise liability, 31
Entities securities coverage. *See* Side C coverage
Environmental liability coverage
  claims-made and reported trigger in, 154
  extended reporting period for, 155
  implications for, 6–7, 143–144
  policy period for, 155
Environmental, social, and corporate governance (ESG), 110–111
EPA. *See* U.S. Environmental Protection Agency
Equipment breakdown insurance. *See* Boiler and machinery insurance
Errors and omissions coverage (E&O coverage). *See* Professional liability coverage
ERUs. *See* Emissions-reduction units
ESG. *See* Environmental, social, and corporate governance
EU. *See* European Union
EU ETS. *See* European Union Emission Trading System
European Union Emission Trading System (EU ETS), 120–122
European Union (EU), 36
Exchange Act, 139

Exclusions
  betterment, 143
  in commercial general liability policies, 151
  D&O bodily injury/property damage, 180–181
  expected or intended injury, 180, 181–183
  fines, penalties, and punitive damages, 7, 144, 146
  fines, penalties, and taxes exclusion, 180, 188–189
  injunctive relief, 180, 186
  intentional/criminal acts, 7, 140–141, 143–144, 145, 180, 189–190
  known loss, known injury or damage, loss in progress clauses, 180, 183–185
  known pollution event, 180
  pollution, 40, 140, 142, 143, 144
  preexisting condition, 180, 185–186
  product liability, 146
  punitive, treble, exemplary, or multiple damages, 180, 187–188
Expected or intended injury exclusion, 180, 181–183
Export Import Bank, 116
Exposure drivers
  claims-made and reported, 154–156
  considerations, 194–195
  limitations, 191–193
  notice requirements, 193–194
  occurrence-based policies, 156–164
  pollution exclusions and grants of pollution coverage, 164–190
Exposure trigger theory, 161
Extended reporting period coverage, 155
Exxon Mobil Corporation, 101–102. *See also Kivalina v. ExxonMobil Corp; Native Village of Kivalina v. ExxonMobil Corporation*

F

Failure to warn claims, 142
False advertising, 123–128
False Advertising Law, 126
Federal common-law nuisance claims, 51
Federal displacement doctrine, 79–80

Federal Trade Commission (FTC)
  focus on "green" marketing, 128–129
  Green Guides, 123, 128–129
Fiji bottled water, 126
Fiji Water Co., 126
Fines, penalties, and punitive damages exclusion, 7, 144
Fines, penalties, and taxes exclusion, 180, 188–189
First-party coverage
  contractors protective professional indemnity and liability insurance, 145
  implications for property coverage, 136
  implications for property (time element) coverage, 136–138
  owners protective professional indemnity insurance, 145
Fluorinated ethers
  exclusion from definition of air pollution, 41
  GHG emissions reports for, 26
  as source targeted for regulation, 16, 19
Ford Motor Company, 80
Fossil-fuel-fired electric power plants, 51, 60
Friends of the Earth, Inc. v. Mosbacher, 116
FTC. *See* Federal Trade Commission
FTSE4Good Index Series (FTSE4Good), 2, 112–113

G

GAAP. *See* Generally Accepted Accounting Principles
GE. *See* General Electric Co.
General Electric Co. (GE), 100
General liability policies, 171
General Motors Corporation, 80
Generally Accepted Accounting Principles (GAAP), 106
Gerrard, Michael B., 91
GHGs. *See* Greenhouse gases
Gidumal v. Site 16/17 Development, LLC, 132–133
Gifford, Henry, 133
Gifford v. U.S. Green Building Council, 133
Global warming, definition in the insurance context, 13–14
Global Warming Solutions Act (GWSA), 58
Goodwyn, Bernard, 157

Green building
  commercial general liability policies and, 142
  contractual and other disputes, 132–134
  litigation, 129–134
  litigation from governmental legislation and regulation, 131
  professional liability coverage and, 144–145
  standards, 130–131
Green Climate Fund, 37
Green Guides, 123
Greenhouse gases (GHGs)
  as air pollutants, 40
  California Environmental Quality Act and, 118
  California legislation regulating, 57–59
  causation issue, 14–16
  claims for natural resource damages pursuant to CERCLA and, 96
  Climate Registry reporting protocol for, 114–115
  considered as pollutants, 165–167
  designing exclusions and coverage for, 17
  emitters of, 26
  EPA Mandatory GHG Reporting Rule, 26, 27–30, 45–46, 115
  exposure trigger theory and, 161
  injury-in-fact trigger theory and, 163
  markets, 120–122
  mobile sources of, 28
  not as pollutants, 169
  as pollutants for purposes of pollution exclusion, 167–170
  potential GHG-related enforcement actions, 134
  reduction agreements between states, 59–61
  regulation under Prevention of Significant Deterioration Program, 46–50
  regulation under Tailoring Rule, 46–50
  shareholder interest in climate change and, 102
  sources targeted for regulation, 16–17
  Title V programs, 46–50
  use of Clean Air Act to regulate emissions, 38–43
Greenwashing
  commercial general liability policies and, 127, 142

consumer actions related to misleading or false advertising and, 123–128
definition of, 122–123
Federal Trade Commission focus on "green" marketing, 128–129
litigation, 122–129
Gross negligence standard, 147
GWSA. *See* Global Warming Solutions Act

H

H-Galdens, 19
H$_2$O. *See* Water vapor
Halons, 41
Harleysville Mutual Insurance Co. v. Buzz Off Insect Shield, LLC, 127
Hazardous substances, 96
HCFEs. *See* Fluorinated ethers
Health coverage, 8
HFEs. *See* Hydrofluorinated ethers
Hill v. Roll International Corp., 126
Honda Motor Company Ltd., 80, 123–125
Hurricane Katrina, 25, 82, 84, 86, 91
Hydrofluorinated ethers (HFEs)
 in definition of air pollution, 41
 GHG emissions reports for, 45
 GHG emissions reports for emissions of, 26
Hydrofluorocarbons (HFCs)
 Cause or Contribute Finding for, 167
 Clean Air Act emission thresholds for, 168
 Climate Registry reporting protocol for, 114
 in definition of air pollutant, 40
 Endangerment Finding for, 167
 GHG emissions reports for, 45, 167
 light-duty vehicle emission standards for, 167, 168
 percentage of U.S. emissions, 17
 as source targeted for regulation, 16, 19
Hydrogen sulfide, 165

I

IDACORP, Inc., 101
IEA. *See* International Energy Agency
IGCC units. *See* Integrated gasification combined cycle units
In re Katrina Canal Breaches Consolidated Litigation, 91

India
 Durban Platform for Enhanced Action and, 38
 Kyoto Protocol and, 34
Injection wells, 52
Injunctive relief exclusions, 180, 186
Injury-in-fact trigger theory, 162–163
Insurance coverage litigation, 92–93
Integrated gasification combined cycle (IGCC) units, 50
Intentional/criminal acts exclusions, 7, 140–141, 142, 143–144, 145, 180, 189–190
Intergovernmental Panel on Climate Change (IPCC)
 attribution defined by, 15–16
 climate change defined by, 14
 EPA Endangerment Finding and, 41, 42
 Fourth Assessment Report, 83
 report on climate change, 11–12
International Center for Technology Assessment, 116
International Energy Agency (IEA), 53–54
International Garment Technologies, LLC, 127–128
International Organization for Standardization (ISO), 148, 171
Inventory of U.S. Greenhouse Gas Emissions and Sinks: 1990–2009, 17, 27
IPCC. *See* Intergovernmental Panel on Climate Change
ISO. *See* International Organization for Standardization
ISO Limited Exception for Short-Term Pollution Events, 177

J

Jackson, Lisa, 40, 42
JI. *See* Joint implementation
Joint implementation (JI), 35

K

Kerry-Boxer Bill. *See* Clean Energy Jobs & American Power Act
Kerry, John, 51
Kivalina v. ExxonMobil Corp., 3, 25
 in *AES Corp. v. Steadfast Insurance Co.*, 169

Kivalina v. ExxonMobil Corp., (continued)
    claims for natural resource damages pursuant to CERCLA and, 96
    commercial general liability policies and, 142
    defense costs in, 191
    directors and officers coverage and, 141
    environmental liability coverage and, 143
    injury-in-fact trigger theory and, 163
    limited pollution endorsements and, 177
    long-tail claims and, 164
    manifestation trigger theory and, 162
    overview of, 76–80
    pollution exclusions and, 176
    Steadfast Insurance Co. and, 92, 156–158
    use of annual reports to support their allegations, 115
Kmart Corporation, 128
Known loss, known injury or damage, loss in progress clauses, 180, 183–185
Known pollution event exclusion, 180
Koh v. SC Johnson & Son, Inc., 125
Kyoto Protocol, 34–35, 37, 114

L

Lanham Act, 133
LEED Green Building Rating System, 30, 130–133, 144
Levin, Bruce A., 22
Liability
    claims arising out of corporate disclosure, 97–115
    disclosure, 102
    enterprise, 31
    limitation of, 7, 140
    potential claims, 31
Liberty Mutual, 138
Light-duty vehicle GHG standards, 43
Limitations
    anti-stacking clauses, 192–193
    batch clauses/multiple claim clauses, 191–192
    defense within or outside of limits, 191
Limited pollution coverage endorsements, 142, 177–178

Litigation. *See also* specific cases; Tort litigation
    arising out of efforts to address climate change issues, 115–134
    asbestos, 22
    carbon capture and sequestration, 52–53
    causation issue, 88–90
    claims arising out of corporate disclosure and management of climate change risk, 97–115
    claims for natural resource damages pursuant to CERCLA, 95–97
    disputes arising out of carbon and GHG markets, 120–122
    foreseeability, 91–92
    green building, 129–134
    greenwashing, 122–129
    insurance coverage, 92–93
    likely claimants, 25–26
    misrepresentation, concealment, or mismanagement of climate change risk, 102
    potential GHG-related enforcement actions, 134
    potential impacts, 25–26
    products liability claims, 87–88
    public trust, 93–95
    resulting from governmental legislation and regulation, 131
    targets/defendants, 26–31
    tobacco, 22
    tort, 70–92
    use of National Environmental Policy Act to address climate change, 115–117
    use of state environmental policy acts to address climate change, 117–120
Lockabey v. American Honda Motor Co., 125
Long-tail claims, 164, 193
Loss-control services, 148

M

Maintenance exclusions, 143, 187
Manifestation trigger theory, 161–162
Maples, Gerald, 22, 87
Market share liability theory, 88–90
Martin Act of 1921, 106–108, 139
Maryland, 132

Massachusetts v. EPA, 38, 39–40, 73–75, 84–85, 92, 167, 168
MATS. *See* Mercury and air toxics standards
Mercury and air toxics standards (MATS), 55
Methane (CH$_4$)
    Clean Air Act emission thresholds for, 168
    Climate Registry reporting protocol for, 114
    in definition of air pollutant, 40
    in definition of air pollution, 41
    Endangerment Finding for, 167
    GHG emissions reports for, 26, 167
    light-duty vehicle emission standards for, 167, 168
    percentage of U.S. emissions, 17
    as source targeted for regulation, 16, 18
Mississippi River Gulf Outlet (MRGO), 91
Mitigation, 1
Montana Supreme Court, 94
Montreal Protocol on Substances that Deplete the Ozone Layer, The, 19
MRGO. *See* Mississippi River Gulf Outlet
Multiple claim clauses, 191–192

N

N$_2$O. *See* Nitrous oxide
NAAQS. *See* National Ambient Air Quality Standards
NAIC. *See* National Association of Insurance Commissioners
National Ambient Air Quality Standards (NAAQS), 47
National Association of Insurance Commissioners (NAIC)
    California, New York, and Washington mandatory survey, 66
    Ceres report on 2010 NAIC survey results, 63–65
    Insurer Climate Risk Disclosure Survey, 62–63, 147
    recent activities, 65–66
    Task Force white paper, 61–62
National Automobile Dealers Association, 58
National Environmental Policy Act of 1969 (NEPA), 115–117
National Highway Traffic Safety Administration (NHTSA), 43, 116
National Research Council (NRC), 11, 41, 42
National Resource Defense Council, 116
National Union Fire Insurance Co. of Pittsburgh, Pa. v. U.S. Liquids, Inc., 179
Native Village of Kivalina v. ExxonMobil Corporation, 79
Natural gas combined cycle (NGCC) power plants, 51
Natural resource damages (NRD)
    claims pursuant to CERCLA, 6, 95–97
    commercial general liability policies and, 142
    environmental liability coverage and, 7, 143
Natural Resources Agency, 118
NEPA. *See* National Environmental Policy Act
New Mexico, 131
New Source Review (NSR), 47
New York, 106–108, 126
New York Executive Law § 63(12), 107
NF$_3$. *See* Nitrogen trifluoride
NGCC power plants. *See* Natural gas combined cycle power plants
NHTSA. *See* National Highway Traffic Safety Administration
Nissan Motor Company Ltd., 80
Nitrogen dioxide, 165
Nitrogen trifluoride (NF$_3$)
    exclusion from definition of air pollution, 41
    GHG emissions reports for, 45
    GHG emissions reports for emissions of, 26
    as source targeted for regulation, 16, 19
Nitrous oxide (N$_2$O)
    Cause or Contribute Finding for, 167
    Clean Air Act emission thresholds for, 168
    Climate Registry reporting protocol for, 114
    in definition of air pollutant, 40
    in definition of air pollution, 41
    Endangerment Finding for, 167
    GHG emissions reports for, 26, 45, 167
    light-duty vehicle emission standards for, 167, 168
    percentage of U.S. emissions, 17
    as source targeted for regulation, 16, 18

No More Excuses Energy Act, 38
Noncumulation clauses, 192
Notice requirements, 193–194
NRC. *See* National Research Council
NRD. *See* Natural resource damages
NSR. *See* New Source Review

O

O_3. *See* Ozone
Obama, Barack, 58, 81
Occurrence-based policies
    AES Corp. v. Steadfast Insurance Co., 156–158
    Cinergy Corp. v. Associated Electric & Gas Insurance Services, Ltd., 159–160
    effect of the "trigger" on occurrence, 160–164
    overview of, 156
OPPI. *See* Owners protective professional indemnity insurance
Our Children's Trust, 93
Overseas Private Investment Corp., 116
Owens Corning v. National Union Fire Insurance Co. of Pittsburgh, Pa., 179
Owners protective professional indemnity insurance (OPPI), 145–146
Ozone (O_3), 17, 41

P

Paduano v. American Honda Motor Co., 123–125
Pay-As-You-Drive (PAYD), 67
PAYD. *See* Pay-As-You-Drive
Peabody Energy, Inc., 107
Perfluorocarbons (PFCs)
    Cause or Contribute Finding for, 167
    Clean Air Act emission thresholds for, 168
    Climate Registry reporting protocol for, 114
    in definition of air pollution, 41
    Endangerment Finding for, 167
    GHG emissions reports for, 26, 45, 167
    percentage of U.S. emissions, 17
    as source targeted for regulation, 16, 18
Peters, Heather, 125

PFCs. *See* Perfluorocarbons
Policy period coverage, 155
Political question doctrine, 71–72
Political risk coverage, 8
Pollutants, 167–170
    as defined by Clean Air Act vs. insurance policy definitions of, 168
    definition of, 164–165
    EPA definition of air pollution and, 41
    interpretation of the term, 165–167
    pollution exclusion for, 40
    U.S. Supreme Court ruling on, 40
Pollution coverage, grants of, 179–180
Pollution exclusions
    absolute, 172–175
    in commercial general liability policies, 142, 168, 170–176
    in directors and officers coverage, 140
    in environmental liability coverage, 143
    grants of pollution coverage and, 164–180
    for pollutants, 40
    in professional liability coverage, 144
    qualified, 171–172
    total, 175–176
    types of, 171–176
Potential claims liability, 9, 31, 135–136
Potential enterprise liability, 8
    based on corporate disclosures, 147–148
    related to provision of loss control consulting services, 148
Preexisting condition exclusions, 180, 185–186
Presidential Memorandum, 58
Prevention of Significant Deterioration Program (PSD), 44, 47
PRI. *See* Principles for Responsible Investment
Principles for Responsible Investment (PRI), 110–111
Pro, Philip, 80
Procter & Gamble Co., 126
Product development strategies, 2, 8–9
Product liability exclusions, 146
Product liability insurance, 151, 191
Products liability claims, 87–88, 142

Professional liability coverage (E&O coverage)
  batch clauses in, 191
  claims-made and reported trigger in, 154
  defense within or outside of limits in, 191
  extended reporting period for, 155
  implications for, 7, 144–146
  policy period for, 155
Property coverage, 8, 136, 149–150
Property (time element) coverage
  implications for, 136–138
  implications for business interruption coverage, 137
  implications for contingent business interruption coverage, 137
  implications for delay in completion coverage under builders-risk policies, 137–138
  implications for service interruption coverage, 137
PSD. *See* Prevention of Significant Deterioration Program
Public trust doctrine, 93–95
Public trust litigation, 93–95
Punitive, treble, exemplary, or multiple damages exclusion, 180, 187–188

Q
Qualified pollution exclusion, 171–172
Quebec, 120

R
RCRA. *See* Resource Recovery and Conservation Act
Reckless disregard standard, 147
Regional Greenhouse Gas Initiative (RGGI), 54, 60–61
Regulations
  congressional action, 51–52
  impact on climate change related litigation, 51
  international backdrop, 34–38
  local, 56–59
  national, 38–51
  potential litigation related to carbon storage and, 52–53
  regional agreements, 59–61
  renewable energy, 53–56
  state, 56–59
  state insurance, 61–66
Renewable energy
  boiler and machinery insurance, 150
  challenges for projects, 55–56
  commercial general liability insurance, 151
  insurance policies tailored for industry, 151–152
  overview of, 53–56
  property insurance policies and, 149–150
  traditional insurance for products, 149–151
Renewable portfolio standards (RPS), 54
Resource Recovery and Conservation Act (RCRA), 169
Retroactive date requirement, 156
RGGI. *See* Regional Greenhouse Gas Initiative
Risk management, 2, 8–9
Risk-management consultants, 148
Roll International Corp., 126
RPS. *See* Renewable portfolio standards

S
Safe Drinking Water Act, 52
SAM. *See* Strategic Asset Management USA, Inc.
Sarbanes-Oxley Act, 106
SC Johnson & Son, Inc., 125–126, 127–128
SEC. *See* U.S. Securities and Exchange Commission
Section 1603 grant program, 54, 56
Section 21(a) reports, 139
Securities Act of 1933, 104
Securities Exchange Act of 1934, 104
Self-insured retention (SIR), 145
SEPAs. *See* State environmental policy acts
Service interruption coverage, 137
Service providers, 30–31
SF_6. *See* Sulfur hexafluoride
Shareholder resolutions, 5, 98–102, 138, 140
Short-term pollution events, 177–178
Side A coverage, 139n9
Side B coverage, 139n9
Side C coverage, 139n9
Sierra Club, 116
Sindell v. Abbott Laboratories, 89

SIPs. *See* State implementation plans
SIR. *See* Self-insured retention
Solar energy, 53
Solar panel warranty insurance, 151–152
Solar photovoltaic output insurance, 152
State environmental policy acts (SEPAs)
    California Environmental Quality Act, 117–119
    implications, 119–120
    property (time element) coverage and, 136
    use of to address climate change, 117–120
State implementation plans (SIPs), 48
State insurance regulation
    California Green Insurance Act of 2010, 66–67
    National Association of Insurance Commissioners, 61–67
    Pay-As-You-Drive regulations, 67
State reporting programs, 46
Strategic Asset Management USA, Inc. (SAM), 111–112
Styrene, 165
Sulfur dioxide, 46
Sulfur hexafluoride (SF_6)
    Cause or Contribute Finding for, 167
    Clean Air Act emission thresholds for, 168
    Climate Registry reporting protocol for, 114
    in definition of air pollution, 41
    Endangerment Finding for, 167
    GHG emissions reports for, 26, 45, 167
    percentage of U.S. emissions, 17
    as source targeted for regulation, 16, 18–19
Superfund. *See* Comprehensive Environmental Response, Compensation, and Liability Act

T
Tailoring Rule, 46–50
Tailpipe Rule, 43–44
Tender Corp., 128
TerraChoice, 127
Texas Commission on Environmental Quality, 94
Third-party coverage
    commercial general liability coverage, 141–142
    contractors protective professional indemnity and liability insurance, 145–146
    D&O coverage, 4–5, 138–141
    environmental liability coverage, 143–144
    owners protective professional indemnity insurance, 145–146
    potential for bad faith claims, 146–147
    professional liability coverage, 144–145
Thomas, Clarence, 74
Thomas, Sidney, 80
Tide, 126
Timing Rule, 44
Title V, Clean Air Act, 44, 47
Tobacco litigation, 22
Tobacco settlement, 89
Tort litigation. *See also* Litigation; specific cases
    climate change related, 3, 70–81
    future trends, 87–92
    related to episodic climatic events, 81–87
Total pollution exclusion, 175–176
Toyota Motor Corporation, 80, 125
Trigger-of-coverage theories
    continuous trigger theory, 163–164
    effect of the trigger on occurrence in, 160–161
    exposure trigger theory, 161
    injury-in-fact trigger theory, 162–163
    manifestation trigger theory, 161–162
Triple trigger theory. *See* Continuous trigger theory
True v. American Honda Motor Co., 124
Turner v. Murphy Oil USA, Inc., 3, 81–82, 177

U
UIC Program. *See* Underground Injection Control (UIC) Progarm
Umbrella/excess policies, 154, 191
Underground Injection Control (UIC) Program, 52
UNEP. *See* United Nations Environment Programme
UNEP FI. *See* United Nations Environment Programme Finance Initiative
Unfair Competition Law, 125, 126

UNFCCC. *See* United Nations Framework Convention on Climate Change
United Nations, 121
United Nations Environment Programme Finance Initiative (UNEP FI), 65–66
United Nations Environment Programme (UNEP), 12
United Nations Framework Convention on Climate Change (UNFCCC)
  Bali Action Plan, 37
  Berlin Mandate, 34
  goal of, 34
  Kyoto Protocol, 34
  recent negotiations, 36
United States
  Durban Platform for Enhanced Action and, 38
  Kyoto Protocol and, 34, 36
U.S. Army Corps of Engineers, 91
U.S. Chamber of Commerce, 58
U.S. Conference of Mayors Climate Protection Agreement, 56–57
U.S. Court of Appeals
  D.C. Circuit, 58
  Fifth Circuit, 84–85, 86, 87, 91, 179
  Ninth Circuit, 79, 81
  Second Circuit, 72, 74, 75
  Sixth Circuit, 179
U.S. Department of Energy (DOE), 54
U.S. District Court
  Central District of California, 124
  District of Columbia, 95
  District of Nevada, 80
  Eastern District of California, 57–58
  Northern District of California, 95, 125
  Southern District of Mississippi, 86
  Southern District of New York, 71, 133, 162
U.S. Energy Information Administration, 53
U.S. Environmental Protection Agency (EPA)
  action pursuant to the Clean Air Act, 38–43
  air pollution defined by, 41
  Cause or Contribute Finding, 40–41
  Endangerment Finding, 40–43
  impact of regulations on demand for renewable energy, 55
  Inventory of U.S. Greenhouse Gas Emissions and Sinks, 17, 27
  light-duty vehicle GHG standards, 43–44
  Mandatory GHG Reporting Rule, 26, 27–30, 45–46, 115, 144, 167
  Massachusetts v. EPA, 38, 39–40, 73–75, 84–85, 92
  mercury and air toxics standards, 55
  rules for carbon dioxide emissions from electric utility generating units, 50–51
  Tailoring Rule, 46–50
  Tailpipe Rule, 43–44
  Timing Rule, 44
  use of CO_2 equivalent to compare roles of GHGs, 97
U.S. Global Change Research Program (USGCRP), 41, 42
U.S. Green Building Council (USGBC), 30, 130–131
U.S. House of Representatives, 38, 51
U.S. Securities and Exchange Commission (SEC)
  Ceres September 2007 petition, 103
  Commission Guidance Regarding Disclosure Related to Climate Change, 98, 104, 105
  guidance on climate change disclosures, 5
  June 12, 2009 petition, 103–104
  petitions for further regulation of climate change risk disclosure, 102
  Regulation S-K, 105
  Regulation S-K, Item 101, 105
  Regulation S-K, Item 103, 105
  Regulation S-K, Item 303, 105–106
  Regulation S-K, Item 503, 105
  role in climate risk disclosure, 102–106
  Rule 10b-5, 104
  Rule 14a-8(i), 100
  Rule 14a-8(i)(7), 100–101
  section 21(a) report, 139
  types of enforcement actions by, 138–139
U.S. Senate, 38

U.S. Supreme Court
   American Electric Power Co., Inc. v. Connecticut, 3, 23n7, 25, 26, 51, 70, 71–76, 79, 143, 176, 191
   Massachusetts v. EPA, 38, 39–40, 73–75, 84–85, 92, 167, 168
USGBC. *See* U.S. Green Building Council
USGCRP. *See* U.S. Global Change Research Program

V

Virginia Supreme Court, 6, 92, 141, 157, 158, 168–169
Virginia Trial Lawyers Association, 157

W

Warranty policies, 8
Water vapor (H$_2$O)
   in *AES Corp. v. Steadfast Insurance Co.*, 169
   exclusion from definition of air pollution, 41
   as source targeted for regulation, 17, 19–20
Waxman-Markey Bill. *See* American Clean Energy and Security Act
WBCSD. *See* World Business Council for Sustainable Development
WCI. *See* Western Climate Initiative
Western Climate Initiative (WCI), 46, 61
Windex, 125
WMO. *See* World Meteorological Organization
Women's Voices for the Earth, Inc. v. Procter & Gamble Co., 126
World Business Council for Sustainable Development (WBCSD), 114
World Business Summit on Climate Change, 110
World Economic Forum, 109–110
World Meteorological Organization (WMO), 12
World Resources Institute (WRI), 114
WRI. *See* World Resources Institute

X

Xcel Energy, Inc., 107

Z

Zurich American Insurance Company, 138